CAMBRIDGE COUNTY GEOGRAPHIES

General Editor: F. H. H. GUILLEMARD, M.A., M.D.

CORNWALL

Cambridge County Geographies

CORNWALL

by

S. BARING-GOULD

With Maps, Diagrams, and Illustrations

Cambridge:

at the University Press

1910

CAMBRIDGE UNIVERSITY PRESS
Cambridge, New York, Melbourne, Madrid, Cape Town,
Singapore, São Paulo, Delhi, Mexico City

Cambridge University Press
The Edinburgh Building, Cambridge CB2 8RU, UK

Published in the United States of America by Cambridge University Press, New York

www.cambridge.org
Information on this title: www.cambridge.org/9781107612341

First published 1910
First paperback edition 2013

A catalogue record for this publication is available from the British Library

ISBN 978-1-107-61234-1 Paperback

CONTENTS

ILLUSTRATIONS

ILLUSTRATIONS

The illustrations on pp. 3, 4, 9, 11, 12, 13, 15, 16, 24, 32, 36, 37, 39, 41, 42, 47, 48, 49, 50, 54, 59, 60, 70, 73, 79, 81, 84, 86, 89, 91, 93, 95, 99, 105, 112, 113, 114, 117, 118, 120, 121, 123, 127, 138, 150, 151, 152, 153, 156, 158 are from photographs supplied by Messrs F. Frith & Co.; those on pp. 8, 43, 45, 58, 77, 94, 108, 109, 134 by Messrs Preston, Penzance; those on pp. 141, 146 by Mr Emery Walker, and those on pp. 6, 29, 34, 87, 119, 124, 130, 132 by Messrs Hayman & Son, Launceston.

1. County and Shire. Meaning of the word.

If we take a map of England and contrast it with a map of the United States, perhaps one of the first things we shall notice is the dissimilarity of the arbitrary divisions of land of which the countries are composed. In America the rigidly straight boundaries and rectangular shape of the majority of the States strike the eye at once; in England our wonder is rather how the boundaries have come to be so tortuous and complicated—to such a degree, indeed, that until recently many counties had outlying islands, as it were, within their neighbours' territory. We may guess at once that the conditions under which the divisions arose cannot have been the same, and that while in America these formal square blocks of land, like vast allotment gardens, were probably the creation of a central authority, and portioned off much about the same time; the divisions we find in England own no such simple origin. Our guess would not have been wrong, for such, in fact, is more or less the case. The formation of the English counties in many instances was (and is—for they have altered up to to-day) an affair of slow growth.

King Alfred is credited with having made them, but inaccurately, for some existed before his time, others not till long after his death, and their origin was—as their names tell us—of very diverse nature.

Let us turn once more to our map of England. Collectively, we call all our divisions counties, but not every one of them is accurately thus described. Cornwall, as we shall see, is not. Some have names complete in themselves, such as Kent and Sussex, and we find these to be old English kingdoms with but little alteration either in their boundaries or their names. To others the terminal *shire* is appended, which tells us that they were *shorn* from a larger domain—*shares* of Mercia or Northumbria or some other of the great English kingdoms. The term county is of Norman introduction,—the domain of a *Comte* or Count.

Although we use the term county for Cornwall, we should not in accuracy do so, as just stated, for it is a Duchy, and has been such since March 17, 1337, when Edward of Woodstock, eldest son of King Edward III, was created Duke of Cornwall. Nor can it be called a shire, for Cornwall was a territory to itself. In 835 Athelstan drove the Britons across the Tamar and made that river the boundary between the Briton and the West Saxon of Devon.

The ancient name of Cornwall and Devon was Totnes, i.e. *Dod-ynys*, "the projecting island," and the Celtic population was that of the Dumnonii. It was not till the tenth century that the name Cornweales appears, signifying the Welsh of the Horn of Britain. The

Latin form of Cornwall is Cornubia. The ancient British settlers in the present department of Finistère called that portion of Gaul Cornouaille.

2. General Characteristics.

On many accounts Cornwall may be regarded as one of the most interesting counties of England, whether we

Luxulyan Village

regard it for its coast scenery, its products, or its antiquities. It has lain so much out of the main current of the life of England that it was hardly mixed up with the politics of the nation till the time of the Civil War.

Its situation, projecting as it does into the sea, by which it is washed on all sides but one, has naturally

caused the natives to take to the water, and has made
Cornwall to be the mother of a hardy breed of fishermen
and sailors. But the county being also rich in mineral
wealth has from an early age caused a large portion of the
manhood of the land to seek their livelihood in mines ;
and the peculiar conditions of Cornwall have thus deter-

Dozmare Pool

mined the professions of a large proportion of its males to
be either on the water or under ground.

The interior of the county cannot be regarded as
beautiful, consisting of a backbone of elevated land, wind-
swept, and over a large area covered with mine-ramps and
the skeletons of abandoned machine-houses standing up
gaunt amidst the desolation. But the valleys are always
beautiful, and the Bodmin moors, if not so lofty and broken

as Dartmoor, are yet fine, and Brown Willy, Rough Tor, and Kilmar are really noble tors.

On the Bodmin moors is Dozmare Pool, the only lake, excepting Loe Pool, that exists in Cornwall. It is small and shallow. There were others formerly, now encroached on or smothered by morass.

In Cornwall it is quite possible to take a stride from the richest vegetation into the abomination of desolation. It has been said in mockery that Cornwall does not grow wood enough to make coffins for the people. The old timber was cut down to supply the furnaces for smelting tin, and it is true that there is not in Cornwall as magnificent timber as may be seen in other counties, but the valleys are everywhere well wooded, and the Cornish elm, that grows almost like a trimmed poplar, stands up lank above the lower trees and coppice.

3. Size, Shape, Boundaries.

Cornwall bears a certain resemblance to Italy, each is like a leg or boot, but Italy stands a-tiptoe to the south, whereas Cornwall is thrust out to the west. But, whereas Italy is kicking Sicily as a football, Cornwall has but the shattered group of the Scilly Isles at its toe.

It touches but one other county, Devonshire, on the east; on all other sides it is washed by the sea, the Atlantic on the north and the English Channel on the south. The heel is the curious projection of the Lizard, and the toe is Land's End. On the east the river Tamar forms mainly the boundary between itself and Devon, except just

The Tamar, near Calstock

north of Launceston, where a small portion of Devonshire juts into Cornwall, bounded on the south by the river Attery, and comprising the parishes of North Petherwin and Werrington. This is due to the land in these parishes having belonged to the Abbey of Tavistock, and the monks desiring to have all their lands comprised in one county. The area of Cornwall is 886,384 acres, or 1385 square miles.

It is the most westerly county in England, and also the most southerly. Its greatest length from the N.E. corner beyond Morwenstow to the Land's End is 80 miles; and its greatest breadth between Marsland Mouth and Rame Head is 46 miles. But it shrinks towards the toe, and between St Ives' Bay and Mounts Bay it is not five miles across.

The Scilly Isles, situated twenty-five miles S.W. from the Land's End, are a part of Cornwall, and have an area of 4041 acres. Formerly, a part of the township of Bridgerule, with 1010 acres on the Devon side of the Tamar, belonged to Cornwall, but has now been dissevered and annexed to Devonshire.

The north coast is sadly deficient in harbours. Bude Haven can accommodate only the smallest vessels, Boscastle is a dangerous creek, Padstow Harbour is barred by the Doom Bank lying across the entrance, and there is none other till we reach St Ives' Bay. On the south coast are Mounts Bay, Falmouth, Charlestown in St Austell Bay, Par, Fowey, Looe, Cawsand Bay, and the Hamoaze that opens into Plymouth Sound. Of these only Falmouth Harbour, once the great station for the packet boats, is good.

The Scilly Isles comprise 145 rocky masses, six only are large islands, and five only are inhabited. The other inhabited islands about Cornwall are very small, these are St Michael's Mount and Looe Island. The promontory of Lleyn in Cardiganshire presents a curious resemblance to Cornwall, and as Cornwall has its detached group of islands in Scilly, so Lleyn has its Bardsey.

Grimsby Channel and Eastern Islands, Scilly

4. Surface and General Features.

To the east and north-east is the large granite mass of the Bodmin moors. It is these striking granitic masses, here and further west—at the Land's End, at St Breage, in the district north of Helston, and again north of St Austell—which form the bolder features of the county. A remarkable depression lies between Marazion and St Ives' Bay, utilised by the railway from Hayle to Marazion road. It almost seems as if the whole of Penwith, the portion west of this trough, had at one time been an island, with a channel of sea between it and the mainland. On the other hand, at a remote

Bodmin

period there can be no doubt that there extended far out broad low-lying lands which are now covered by the sea, for forest beds have been found in Mounts Bay, in Padstow Bay, at St Columb Major, and elsewhere, showing that there has been a subsidence of the land. This has given rise to the fable of a drowned realm of Lyonesse, but this Lyonesse never existed in or near Cornwall; it was Léon in Brittany. But if there has been subsidence, there has also been an elevation of the land, as is shown by the raised beaches that can be traced along the coast. In the south are found flint pebbles in these raised beaches, showing that the wash at one period was not as now from west to east, but the reverse. It was from these flint pebbles in the elevated beaches that prehistoric man in Cornwall obtained the material for the fabrication of his tools and weapons. The elevation of the land, which dried up the channel between the Land's End and the mainland, preceded the depression which sunk the now submerged forests.

To the north of the great granite boss that forms the Bodmin moors a ridge of cold moors rises, setting its back against the Atlantic and feeding the rivers that flow south with the rain that pours over it. Very little can be grown on these heights: they produce a little barley, but are mainly covered with rushes, coarse grass, and furze bushes. No considerable heights are reached till we come to Carnmarth and Carn Brea, each only 700 ft. above the sea. Then there is no distinct height till we reach Godolphin and Tregonning hills of 560 ft. and 600 ft. Towards Land's End there are greater heights—700 ft. and a little over, but this is naught compared with Brown Willy, 1368, and Kilmar, 1297, in the east. But it must be understood that grand mountain, or even fine hill scenery is not to be met with in western Cornwall. Its glory is in its magnificent coast-line; and its beauty is to be found in its lovely valleys and coombes.

The general features of a country depend on its

Rough Tor

geological structure.   Granite formation, slate rock, sandstone, limestone, chalk—all have their special characters,
unmistakable.   When we are among the granites in the
west of England we expect a tent-like shape of hill with
a tor of rock at the summit; the sides strewn with a

St Keyne's Well, Liskeard

"clatter" of fallen rock, and clothed in heather and furze.
When we come to the slates, and the overlying cold clays,
we expect little except in the gorges and valleys cut through
the strata by the streams.   Of sandstone, or of chalk, forming breezy rolling downs, there is none in Cornwall, nor
of limestone with its bold scars, such as are met with in

the western hills of Yorkshire. We must take what we can find—and much can be found in Cornwall if we do not expect too much, nor look for what is not there, and under existing geological conditions could not be there.

5. Watershed, Rivers.

From what has been already said, it will be seen that the great spinal column of Cornwall is in the north and that

On the Camel

consequently the principal rivers must flow to the south. It is true that the Camel and its tributaries rise in the north and flow north to debouch into Padstow estuary, but that is the only river of the smallest consequence that directly feeds the Atlantic.

The Camel rises in the windswept, sodden, clay land
above Boscastle, and dribbles down to Camelford, passing
under Slaughter Bridge and the stone of LATEINOS which
traditionally mark the scene of King Arthur's last fight
and death. After leaving Camelford, it plunges through
a beautiful wooded valley, under Helsbury Castle occupying a bleak conical hill—a castle of the Dukes of Cornwall,
but consisting only of a stone camp of prehistoric date
and a ruined chapel in the midst dedicated to St Sith or
Itha, an Irish saint. It passes Lavethan, a hospitium of
the monks of Bodmin, and between wooded banks through
Pencarrow and Dunmeer woods, and having run south, it
now turns north to meet and mingle waters with the
Allan, which has cleft for itself a very similar and equally
beautiful valley. The Allan rises near the slate quarries of
Delabole, and glides down by St Teath, and by St Mabyn
on its bleak stormbeaten height, itself snug between beautiful
hanging woods and with sweet old-world manor houses
clustering near, and meets the Camel at Egloshayle.
Thence they flow away into the Padstow estuary under
the old 15th century bridge at Wadebridge, past the
Camelot of legend—the only streams of any consequence
that flow north.

With regard to the Tamar, we may call it as we
please a Devonshire or a Cornish river. It divides the
counties through the principal part of its course, but it
has its source in Morwenstow, on a wretched moor, in
Cornwall. Not much can be said in its favour till it
reaches Bridgerule. From that point past North Tamerton
(and Vacey, which although on the left bank was included

in Cornwall, and has two formidable earthworks) it glides down to Werrington, where it meets the waters of the Attery and passes under Polson Bridge within sight of Launceston.   Thenceforth the Tamar is in the full bloom of beauty.   Carthamartha (Caer Tamar) stands at its junction with the Inny.   Below Polson Bridge it has accepted the Lyd from Devon.   Then through the lands

Wadebridge

and woods of the Duke of Bedford at lovely Endsleigh, under the bold crags of Morwell, up to which the tide reaches, then past Calstock and Cothele, and in serpentine writhes about Pentillie Castle, and so into the Hamoaze —the most beautiful river in England, excepting possibly the Wye.

The Inny, one of the feeders of the Tamar and altogether Cornish, must not be omitted, for it is a beautiful stream. It rises in the elevated land by Davidstowe and ripples down near Altarnon, passing in a picturesque valley the Holy Well and chapel of St Clether and the ancestral seat of the Trevelyan family at Basil; then, still in its beautiful valley, past Polyphant, famous for its quarries of a stone that admits of the most delicate carving, until it reaches the Tamar at Innyfoot. It is a river rich in trout.

Royal Albert Bridge, Saltash

An old Cornish song of the Altarnon volunteer has the verse:

O Altarnon! O Altarnon! I ne'er shall see thee more,
Nor hear the sweet bells ringing, nor stand in the church door,
Nor hear the birds a-whistling, nor in the Inny stream
See silver trout glance by me, as thoughts glance by in dream.

It is not however the Inny but a tributary that actually passes Altarnon.

The Lynher falls into the Hamoaze, running for much of its course parallel to the Inny. It rises in the Bodmin moors and flows through the beautiful grounds of Trebartha where it receives a feeder from Trewartha marsh that leaps to meet it in a pretty cascade. Trewartha marsh has been turned over and over for tin and gold, and the Squire of Trebartha formerly furnished his daughters with gold rings made from the precious ore found in it. A curious settlement of the Celtic period exists above the marsh. It has however been much mutilated by farmers, who have carried off the stones for the construction of pigstyes. The Lynher flows through the park of the Earl of St Germans, past the beautiful church with its Norman west front, and then is lost in the united waters of the Tamar and the Tavy.

Under Brown Willy is a pool called Fowey Well, which is traditionally held to be the source of the Fowey river. This however is not the case. It rises under Buttern Hill (1135 ft.) crowned by cairns, but as the Fowey Well has no outlet visible, it is supposed to decant by a subterranean stream into the river. Leaping down from the moors, the Fowey enters a wooded valley and, turning abruptly west, flows through another well timbered valley. Running beside the railway, and then turning sharply south, it passes the old Stannery town of Lostwithiel, to which the tide reaches, and plunging into a narrow glen with St Winnow on one side and Golant on the other, finally reaches the sea at Fowey harbour.

There are two Looe rivers, one rising in the Bodmin moors receives the overflow of Dozmare Pool, and flowing deep below Liskeard receives the West Looe above the estuary. Duloe, which has a small but interesting circle of upright stones, stands between them and is supposed to be so called as between the two Looes. Before reaching Duloe the river has passed under St Keyne, famous for its Holy Well commemorated by Southey in a well-known ballad.

There is no river of any importance till we reach the Fal. Rising on Goss Moor, not far from St Columb, and passing Grampound and Tregony, now an utterly decayed place, it meets the Tressilian and the Truro rivers, and all three, insignificant hitherto, suddenly acquire importance and spread out into the beautiful estuary of the Fal or Carrick Roads. Here are Penryn creek, Mylor creek, and Porthcuel harbour, commanded by the castle of St Mawes. None of these owe their importance to the sweet waters they bring down; all their value is due to the tide that flows up to Truro.

The entrance to the Roads is found between Zoze Point and Pendennis Point, the latter at one time defended by a strong castle. Almost halfway between the points is the dangerous Black Rock, to which a former Trefusis conveyed his wife and there left her to be overwhelmed by the rising tide. Happily she was rescued by some fishermen. The Helford river is but a creek, noted for its oyster beds, into which a little stream dribbles.

6. Geology and Soil.

By Geology we mean the study of the rocks, and we must at the outset explain that the term *rock* is used by the geologist without any reference to the hardness or compactness of the material to which the name is applied; thus he speaks of loose sand as a rock equally with a hard substance like granite.

Rocks are of two kinds, (1) those laid down mostly under water, (2) those due to the action of fire.

The first kind may be compared to sheets of paper one over the other. These sheets are called *beds*, and such beds are usually formed of sand (often containing pebbles), mud or clay, and limestone, or mixtures of these materials. They are laid down as flat or nearly flat sheets, but may afterwards be tilted as the result of movement of the earth's crust, just as you may tilt sheets of paper, folding them into arches and troughs, by compressing their ends. Again, we may find the tops of the folds so produced wasted away as the result of the wearing action of rivers, glaciers, and sea-waves upon them, as you might cut off the tops of the folds of the paper with a pair of shears. This has happened with the ancient beds forming parts of the earth's crust, and we therefore often find them tilted, with the upper parts removed.

The other kinds of rocks are known as igneous rocks, which have been melted under the action of heat and become solid on cooling. When in the molten state they have been poured out at the surface as the lava of volcanoes, or have been forced into other rocks and cooled in the cracks and other places of weakness. Much material is also thrown out of volcanoes as volcanic ash and dust, and is piled up on the sides of the volcano. Such ashy material may be arranged in beds, so that it partakes to some extent of the qualities of the two great rock groups.

The production of beds is of great importance to geologists, for by means of these beds we can classify the rocks according to age. If we take two sheets of paper, and lay one on the top of the other on a table, the upper one has been laid down after the other. Similarly with two beds, the upper is also the newer, and the newer will remain on the top after earth-movements, save in very exceptional cases which need not be regarded by us here, and for general purposes we may regard any bed or set of beds resting on any other in our own country as being the newer bed or set.

The movements which affect beds may occur at different times. One set of beds may be laid down flat, then thrown into folds by movement, the tops of the beds worn off, and another set of beds laid down upon the worn surface of the older beds, the edges of which will abut against the oldest of the new set of flatly deposited beds, which latter may in turn undergo disturbance and renewal of their upper portions.

Again, after the formation of the beds many changes may occur in them. They may become hardened, pebble-beds being changed into conglomerates, sands into sand-stones, muds and clays into mudstones and shales, soft

NAMES OF SYSTEMS		SUBDIVISIONS	CHARACTERS OF ROCKS
TERTIARY	**Recent Pleistocene**	Metal Age Deposits Neolithic „ Palaeolithic „ Glacial „	Superficial Deposits
	Pliocene	Cromer Series Weybourne Crag Chillesford and Norwich Crags Red and Walton Crags Coralline Crag	Sands chiefly
	Miocene	Absent from Britain	
	Eocene	Fluviomarine Beds of Hampshire Bagshot Beds London Clay Oldhaven Beds, Woolwich and Reading Thanet Sands [Groups	Clays and Sands chiefly
SECONDARY	**Cretaceous**	Chalk Upper Greensand and Gault Lower Greensand Weald Clay Hastings Sands	Chalk at top Sandstones, Mud and Clays below
	Jurassic	Purbeck Beds Portland Beds Kimmeridge Clay Corallian Beds Oxford Clay and Kellaways Rock Cornbrash Forest Marble Great Oolite with Stonesfield Slate Inferior Oolite Lias—Upper, Middle, and Lower	Shales, Sandstones and Oolitic Limestones
	Triassic	Rhaetic Keuper Marls Keuper Sandstone Upper Bunter Sandstone Bunter Pebble Beds Lower Bunter Sandstone	Red Sandstones and Marls, Gypsum and Salt
PRIMARY	**Permian**	Magnesian Limestone and Sandstone Marl Slate Lower Permian Sandstone	Red Sandstones and Magnesian Limestone
	Carboniferous	Coal Measures Millstone Grit Mountain Limestone Basal Carboniferous Rocks	Sandstones, Shales and Coals at top Sandstones in middle Limestone and Shales below
	Devonian	Upper Mid Devonian and Old Red Sand- Lower stone	Red Sandstones, Shales, Slates and Limestones
	Silurian	Ludlow Beds Wenlock Beds Llandovery Beds	Sandstones, Shales and Thin Limestones
	Ordovician	Caradoc Beds Llandeilo Beds Arenig Beds	Shales, Slates, Sandstones and Thin Limestones
	Cambrian	Tremadoc Slates Lingula Flags Menevian Beds Harlech Grits and Llanberis Slates	Slates and Sandstones
	Pre-Cambrian	No definite classification yet made	Sandstones, Slates and Volcanic Rocks

deposits of lime into limestone, and loose volcanic ashes into exceedingly hard rocks. They may also become cracked, and the cracks are often very regular, running in two directions at right angles one to the other. Such cracks are known as *joints*, and the joints are very important in affecting the physical geography of a district. Then, as the result of great pressure applied sideways, the rocks may be so changed that they can be split into thin slabs, which usually, though not necessarily, split along planes standing at high angles to the horizontal. Rocks affected in this way are known as *slates*.

If we could flatten out all the beds of England, and arrange them one over the other and bore a shaft through them, we should see them on the sides of the shaft, the newest appearing at the top and the oldest at the bottom, as shown in the table. Such a shaft would have a depth of between 10,000 and 20,000 feet. The strata beds are divided into three great groups called Primary or Palaeozoic, Secondary or Mesozoic, and Tertiary or Cainozoic, and the lowest Primary rocks are the oldest rocks of Britain, which form as it were the foundation stones on which the other rocks rest. These may be spoken of as the Pre-Cambrian rocks. The three great groups are divided into minor divisions known as systems. The names of these systems are arranged in order in the table and on the right hand side the general characters of the rocks of each system are stated.

With these preliminary remarks we may now proceed to a brief account of the geology of the county.

In Cornwall there is a succession of nodes of granite

DIAGRAM SECTION FROM SNOWDON TO HARWICH, ABOUT 200 MILES.

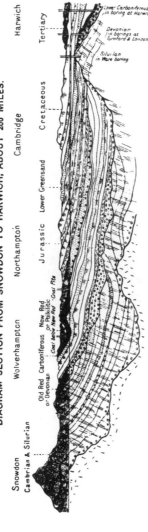

Snowdon

Cambrian & Silurian

Wolverhampton

Old Red Carboniferous New Red
or Devonian   or Poikilitic
         Coal below New Red Coal Pits

Northampton

Jurassic   Lower Greensand

Cambridge

Cretaceous

Harwich

Tertiary

Lower Carboniferous
in boring at Harwich

Devonian
in borings at
Turnford & London

Silurian
in Ware boring

This cross section shows what would be seen in a deep cutting nearly E. and W. across England and Wales. It shows also how, in consequence of the folding of the strata and the cutting off of the uplifted parts, old rocks which should be tens of thousands of feet down are found in borings in East Anglia only 1000 feet or so below the surface.

rising to the surface, a continuation westward of the mass
of Dartmoor.   It has surged to the surface in four large
masses continued westward by the Scilly Isles.   These
granitic masses have upheaved the superincumbent beds of
stratified rocks, partly melting them.   These distinct
nodes are: the Bodmin moors, the St Austell elevation,
the Carn Menelez, and the Land's End district.   Smaller

The Cheesewring

masses of granite occur in the double heights of Godolphin
and Tregonning, St Michael's Mount, Carn Brea and
Carn Marth, and Castel-an-Dinas.

The Elvans are dykes of quartz-porphyry which issue
from the granite into the surrounding slates, and are often
mistakenly supposed to be a bastard granite.

The granite in its upheaval has strangely altered and

contorted the superposed beds. There are as well in-trusive veins of igneous rocks. In the Lizard district is serpentine, a compact, tough rock often of a green colour, lending itself to a high polish, and forming magnificent cliffs with a special gloss and colour, as well as maintaining on the surface a special flora.

The prime feature in Cornish geology is the upheaval of the granite, distorting, folding back, and altering the superincumbent beds.

In the north-east of Cornwall from a line drawn from below Launceston, on the Tamar, to Boscastle the rocks belong to the culm measures of North Devon. All the rest of the peninsula, except the protruding granite and the serpentine of the Lizard, pertains to the Devonian series of sedimentary rocks, in which the first signs of life appear; consisting largely of clay-slate, locally known as Killas, alternating with beds of red or grey grit and sandstone. Although these slaty rocks must be some thousand feet in thickness, they have been so broken up and turned over by the convulsions of the earth that their chronological sequence cannot easily be determined. In these con-vulsions they have been rent, and through the rents have been driven hot blasts that have deposited crystalline veins, or injections of trap and other volcanic matter, altering the character of the rock through which they have been driven. By the Menheniot Station on the G.W.R. is a hill of serpentine thrown up at one jet, and now largely quarried for the sake of the roads.

The culm measures already alluded to consist of black shales and slates with seams of grit and chert, much undulated through enormous lateral pressure. The granite, the lowest and most ancient formation of all, was itself consolidated under vast pressure from above, and was not in a molten condition when forced to the surface. Had it been so, it would have resolved itself into lava. It was cold when upheaved, tearing apart the superincumbent stratified sedimentary rocks, which disappeared from the summits, and on all sides about these upheavals were twisted, contorted, thrown back, and fissured.

Atmospheric effect and natural gravitation is constantly carrying the soil from the upper land, from the hills into the bottoms, and consequently it is in the latter that we find the richest land, best calculated to repay the toil of the agriculturist. On the high moors there is little depth of so called " meat earth," below which is clay and grit, hard and unprofitable, commonly called the " calm " or the "deads." But adjoining the granite is the wash from it of its dissolved felspar, the china-clay that furnishes the inhabitants of the St Austell district with a remunerative and ever-growing industry, of which more presently.

7. Natural History.

Various facts, which can only shortly be mentioned here, go to show that the British Isles have not existed as such, and separated from the Continent, for any great length of geological time. Around our coasts, for instance, and specially in Cornwall, are in several places remains of forests now sunk beneath the sea, and only to be seen at extreme low water. Between England and the Continent the sea is very shallow, and St Paul's Cathedral might be placed anywhere in the North Sea without submerging its summit, but a little west of Ireland we soon come to very deep soundings. Great Britain and Ireland were thus once part of the Continent and are examples of what geologists call recent continental islands. But we also have no less certain proof that at some anterior period they were almost entirely submerged. The fauna and flora thus being destroyed, the land would have to be restocked with animals and plants from the Continent when union again took place, the influx of course coming from the east and south. As, however, it was not long before separation occurred, not all the continental species could establish themselves. We should thus expect to find that the parts in the neighbourhood of the Continent were richer in species and those furthest off poorer, and this proves to be the case both in plants and animals. While Britain has fewer species than France or Belgium, Ireland has still less than Britain.

Small though England may be, she can nevertheless show most striking differences of fauna and flora in different districts. On the moors of the north, for example, the heaths and berries underfoot, and the larger birds of prey and grouse which now and again meet our view offer a marked contrast to—let us say—the furze-clad chalk downs of Sussex, where the wheatear and whinchat and the copper butterflies and "blues" are familiar objects. These differences depend upon a number of conditions, often mutually interdependent—upon variations of soil, rainfall, temperature, and so forth. Cornwall presents unusual peculiarities in many ways, and we may now consider how far these have affected the creatures and plants within her borders.

Firstly, Cornwall is remotely situated—one of the extreme points of Western Europe—and, whether the fact be dependent on food conditions or not, we find that there are several species of bird, common in other parts of England, which do not occur within the county, such as the nightingale, the wood warbler, garden warbler, redstart, and others. It would almost seem as if some of these species had not found their way thither since the re-peopling of the land by its present fauna, but were in gradual process of doing so, for there is no doubt that many birds rare or unknown in the Duchy half a century ago are now not uncommon, and appear to be steadily moving westward. That the starling is doing so is perhaps not remarkable, for this bird has enormously increased in numbers of late years and has spread everywhere, even up into northern Scotland, but it is curious that birds like the stock-dove and all the woodpeckers and other non-gregarious sorts should show this tendency.

Next, Cornwall is from its position constantly exposed to high winds, and to heavy gales in winter, combined with an unusually heavy rainfall and an "insular" climate tending to warmth and equableness. These factors, added to the granitic formation of much of its area, have made it a country of bleak moorland varied with thickly-wooded deep valleys—dampness being the leading characteristic of both. With such physical conditions, then, we should expect to find the Duchy not very varied in its native trees, perhaps, but particularly abundant in ferns, and this is the case, for 39 species are recorded, while lichens are not less rich. It bears in many ways a resemblance to the climate of Portugal, for here the camellia flourishes and displays its beautiful flowers to perfection, and the tea

In a Cornish Garden

plant does so well that there seems no reason why it should not be grown for profit. It is not a land of warblers, nor can it show the rich and varied wildfowl fauna of the Fenlands, but there is no county in England where, in the marshy glens, woodcocks are more abundant. The moorlands, too, abound in snipe, and at one time blackgame were common, but the larger birds of prey have for the most part vanished, though an occasional buzzard may be seen and the raven is not yet extinct.

Lastly, it is to be noted that Cornwall is the nearest part of England to America. However difficult it may be of explanation, the fact remains that the Duchy is very rich in rare birds; so rich, indeed, that their recorded occurrence cannot by any possibility be merely accidental. Thus, no less than 24 species have occurred in Cornwall which have never been found in Devonshire. But more than this, a very large number of these—18 or more—are purely American species. The question is, whence do they come? Professor James Clark, who has discussed the point at some length in the *Victoria County History*, is, apparently, loth to believe that they can come directly across the Atlantic, and it is by many thought that they are driven back by heavy south-westerly weather when dropping down the English Channel, having come by a circuitous route from Northern Europe. But against this is the undeniable fact that it is in the immediate neighbourhood of the Land's End that the chief rarities and stragglers are obtained, while many species have been shot in the Scilly Islands which have never been recorded from Cornwall itself.

So far as its botany is concerned, Cornwall does not differ very markedly from Devonshire, but it has a large number of rare or peculiar plants. The highlands and north coast are rather poor in species; it is on the banks and estuaries of the streams that the richest flora is seen. A number of foreign plants are found, mostly in the neighbourhood of Falmouth and other ports. The balsam,

Impatiens Roylei, from India, grows extremely abundantly between Liskeard and Looe, and near Tintagel, and a species of May-weed (*Matricaria discoidea*) has become a troublesome pest near Falmouth. Loe Pool in the Lizard district is noticeable for the number of rare and local plants it possesses. The Scilly Islands own certain plants peculiar to them ; thus, *Trifolium repens*, var. *Townsendi*, and *Ornithopus ebracteatus* are said not to be found elsewhere in England, and *Carex ligerica* only in Norfolk.

The chief feature of the mammals of the county is that the grey seal, *Halichaerus gryphus*, is quite numerous in the Scilly Islands ; that the polecat, though nearly extinct, it still found ; and that both badgers and otters are very abundant. It is a curious fact that certain freshwater fish common in other parts of England, such as pike, roach, chub, and bream, are unknown.

The bird which bears the distinctive appellation, the Cornish chough (it is not confined to the county, but is also found in Wales), is now not nearly as common as formerly, but like the raven it still breeds on some parts of the coast.

8. Around the Coast. From Morwen= stow to Land's End.

This noble coast—so terrible to sailors—begins with the fine Henna Cliff at Morwenstow. Morwenstow Church contains an early font and has fine Norman arches.

Here is Tonacombe, an interesting early Tudor house quite unspoiled. At Morwenstow lived the Rev. Robert Hawker, a poet and character. Bude Haven is a growing seaside place, with golf-links and tolerable bathing. Stratton, of which parish it actually forms or did form a part, has a fine well-cared for church, and above the town is Stamford Hill, where was fought a battle in the Civil

Bude Breakwater

War, on May 16, 1643. Sir Bevil Grenville and Hopton commanded the Royalist Army, and the Earl of Stamford the Parliamentarians. The latter were defeated with the loss of 300 men killed and 1700 taken prisoners. One of the old guns marks the site, and an inscription in commemoration of the battle is affixed to the Tree Inn. Widemouth Bay has good sands and promises at some future day to become a sea-bathing place superior to Bude. At Dazard the cliffs are fine; at St Gennys is Crackington Cove with a small beach. Beyond this, High Cliff (705 ft.) is reached, the loftiest headland on the coast. The coast is magnificent to Boscastle. Near this is Pentargon, a beautiful bay into which a little stream leaps in a waterfall. Boscastle is a narrow creek into which only in calm weather can small vessels enter. It is sheltered by a headland in which is a blow-hole. In a lovely valley is the towerless church of Minster. In caves about Willapark seals breed. From hence to Tintagel the cliffs are of slate and are quarried, the slate being let down into boats in the water, when weather permits. Before reaching Tintagel we come to St Neighton's (Nectan's) Kieve, a small waterfall in a glen, where maidenhair fern once abounded.

Tintagel village is separated from the church by a deep glen. The church is on a windy height, and is interesting for its antiquity. Tintagel castle stands on a headland, once an island, but the cliff and a portion of the castle have fallen into the narrow gulf and choked it. The sea has bored a tunnel through the headland, and very little of the castle remains. The walls were of the local slate-stone set in mortar made of sea-shells. In this castle, traditionally, King Arthur was born. There are slate quarries in the neighbourhood, and further inland are the Delabole quarries, from which slate is conveyed to all parts of England.

The small Trebarwith Cove is passed and then we reach Port Isaac Bay, which takes its name, not from the

King Arthur's Castle, Tintagel

patriarch, but from a Cornish word that signifies a port for corn.

Porthqueen has its pilchard cellars cut out of the rocks; and at the western end of its bay Pentire Head stands out boldly into the Atlantic with a cliff castle at its extreme point. Pentire to some extent shelters Padstow Bay, but the entrance to the harbour is made dangerous by the Doom Bar lying across it. Here the rounded, sand-powdered Bray Hill is supposed to have buried under the drift sand the remains of a Roman town, but that there was such a Roman settlement is very doubtful. Further inland is the church of St Enodoc recently dug out of the sand, and the path to the porch has singular "Lord's Measures" that have been collected and are here planted. They were measures for grain. Of Padstow something shall be said elsewhere. Before reaching Trevose Head is Harlyn Bay, where from under the sand has been dug out a cemetery of the iron age, the skeletons crouching with their chins to their knees, in slate kists or coffins, with bronze ornaments, glass and amber beads, and iron instruments. Here also by the falling of the earth or sand some gold ornaments of the Celtic period were laid bare. On Trevose Head is a lighthouse. In Constantine Bay are the remains of a church half buried in sand, and with bones strewn about it. The font from this abandoned church has been transferred to St Merryn. It is of black Catacluse stone, as are the piers and arches of the church, giving it a sombre look. Passing the little Porthcothan we come to the noble cliffs of Bodruthan Steps, perhaps the finest bit of the north-coast scenery, and then reach

Bodruthan Steps

Mawgan Porth, the estuary of the tiny stream that waters the vale of Lanherne. Here is the ancient mansion of the Arundells, dating from 1580, granted by Lord Arundell to Carmelite nuns, who came to England, at the outbreak of the French Revolution, in 1794. They have remained here ever since in seclusion. Near the door into their chapel is an old cross with inscription and interlaced work.

Newquay Harbour

The church has a beautifully proportioned tower, a good rood-screen, and many monuments of the Arundells, as well as a remarkable Gothic cross. A grove of Cornish elms like whips occupies the bottom of the valley. The church and parish are St Mawgan in Pyder.

Newquay with its sands is a rapidly growing watering-place, with staring hotels. Crantock church, which once

had its canons, is interesting; it has been carefully restored
and has a fine rood-screen and loft. On the stalls is
carved a dove with a chip of wood in its beak, and the
story goes that St Carantoc—who assisted St Patrick in
drawing up the Senchus Môr, the code of laws for Christian
Ireland—was guided to the site for his church by seeing a
dove carry off the shavings of his staff which he was
cutting. Perran Bay has sent its sands flying inland,
causing a widespread extent of *towans* or dunes, which
have overwhelmed first of all the old church of St Piran
or Cieran of Saighir in Ireland, who was buried in it, and
then a second church built further inland, which also in
turn had to be abandoned. These sand-dunes are a rabbit
warren. The old church of St Piran, extremely rude in
construction, has been excavated, but the only carved work
found has been removed to the museum at Truro. The
church resembles one of the early Christian chapels in
Ireland. Bones are scattered about around it.

St Agnes—or St Aun's Head as it is locally termed—
is but 617 ft. high, but it stands boldly above the sea.

Portreath is a busy little place to which coal-barges
come from Cardiff with coal for the Camborne and Redruth mines. Tehidy Park, which stands above it, is an old
seat of the Bassets. After passing Navax Point the broad
Bay of St Ives opens out, with a lighthouse at each horn,
and here again begin the towans or sand-dunes. Hayle is
at the mouth of a small stream of the same name, and
signifies "saltings." We come now to the town of
St Ives, with an interesting church and an old cross.
St Ives till about 1410 was but a tiny fishing village and

was in the parish of Lelant. Perkin Warbeck landed near here in 1497.

Near this is the Knill monument erected in 1782 by the eccentric John Knill, who left money for an annual procession to it, and a dance of children around it. He intended it to have been his mausoleum, but he died and was buried in London.

The Wharf, St Ives

Zennor has a curious little church with carved bench ends, on one of which is represented a mermaid, and Zennor "quoit" is thought to be the largest cromlech in England.

The Gurnard's Head has sheer cliffs on east and west, composed of slaty felspar, horneblende and greenstone, whereas the rocks of Zennor are of granite and slate in juxtaposition, and with dykes of granite penetrating the slate. Hills of rock and heather and furze bend round in a crescent terminating on one side at Carnminnis, on the other at Carn Galva, enclosing a great basin that reaches to the cliffs. On the isthmus connecting the headland with the mainland is a ruined chapel, with the altar-stone entire. Three miles westward is Morvah with the interesting Chûn Castle above it, of rude stone forming three concentric rings, and not far off is Chûn cromlech.

We come now to Botallack with its famous mine, carried to a depth of a thousand feet and extending a considerable distance under the sea. In time of storm the booming of the waves overhead and the clashing of stones rolled by the billows is so great that the bravest miners are driven from their work. Tin mining is now but languidly carried on. Although the heights are not great, yet this portion of coast is remarkably fine.

Cape Cornwall exhibits the junction of the slate with the granite. Here also extensive mining has been carried on, and Boswedden, like Botallack, burrowed far under the sea. Near the church of St Just is one of the circles or amphitheatres in which miracle plays were performed. The "Merry Maidens" is a prehistoric stone circle in the neighbourhood with ten stones erect and five fallen.

Whitesand Bay enjoys some slight shelter from the S. and E. winds. It is said that this bay was the landing-place of Athelstan after his conquest of Scilly, of King Stephen in 1135, and of King John when he returned from Ireland. In Sennen Cove is a cluster of fishermen's cottages.

Land's End is the end of Penwith, the "chief headland," and the Bolerium of the ancients. It commands a magnificent view, extending to Cape Cornwall over Whitesand Bay, of the Longships rocks with their lighthouse, and in clear weather of the Scilly Isles. The Wolf lighthouse is planted on a dangerous rock of felspathic porphyry some eight miles S.W. from the shore. The Land's End bristles with sharp fangs of rock, and somewhat resembles the back of an alligator.

Land's End

9. Around the Coast. From Land's End to Rame Head.

The south coast-line of Cornwall presents a great contrast to that of the north, except for the portion from the Land's End to Mounts Bay and the Lizard. We have no more the wind-swept background of heights, barren and often tortured by miners, turned into a waste of heaps of rubble and studded with ruined engine-houses. We find instead a gentler sea-board, pierced by long estuaries, and with valleys of rich vegetation running down to the sea. We speedily leave the granite and the culm measures, and are among the rocks of the Devonian series, less stern and forbidding in colour. In St Levan parish

Newlyn Pier

at Trereen Point is the Logan Rock, a block of granite weighing over 65 tons, once so nicely balanced that it could be made to rock by the finger of a child. In 1874 a young naval officer, with the assistance of a boat's crew, upset it. This raised a storm of indignation in Cornwall, and the Admiralty ordered him to replace it, which he did, at great expense.

St Michael's Mount

Mousehole is a village of fishermen and boatmen, and was the residence of Dolly Pentreath, the last person in Cornwall who spoke the Cornish language, as also of some of the unfortunate sailors who joined Captain Allan Gardner in 1850 in his ill-fated missionary expedition to South America. All the members of the mission died of starvation in Tierra del Fuego.

Newlyn has become noted as a place of residence of a school of artists. It is on the wide and beautiful Mounts Bay over against St Michael's Mount.

Penzance ("the Holy Head") has become a great resort of residents for the winter, owing to the mildness of the climate. Marazion or Market Jew does not derive its name from "the Bitter Waters of Zion" as has been absurdly asserted; Marazion has the same significance as Market Jeu (i.e. Jeudi), Thursday Market. Here is a submerged forest. St Michael's Mount is a rock that rises out of the sands and can be reached, when the tide falls, by a causeway. It is incomparably inferior both in elevation and in the dignity of the buildings that crown it to the Mont St Michel in Normandy, but is nevertheless a picturesque adjunct to the scene. Some writers suppose that it is the Ictis to which the natives conveyed the tin and trafficked with the Phoenicians, but it is totally unsuited by nature to serve as a market-site, and there is no certain evidence that the Phoenicians ever came to Cornwall. About Marazion daffodils, narcissus, and violets are cultivated largely for the London market. At Perran Uthnoe the cliffs again appear, and we reach Prussia

Kynance Cove and the Lizard

Cove, once the haunt of smugglers. Inland, Godolphin and Tregonning's granite hills are conspicuous, and near the coast is Pergersick Castle, a picturesque ruin of which strange legends are told. Porthleven is a small fishing village, where the people live on the annual arrival of the pilchards. Loe Pool is a beautiful sheet of water cut off from the sea by a bar of sand. It was when standing on this bar, and watching the wreck of a vessel close in shore when those on the land were unable to communicate with it, that Henry Trengrouse conceived the idea of a rocket apparatus, to be not only employed on land, but also to be carried by every ship. He, of course, met with opposition from the Board of Trade and the Government, and he spent his life and his fortune in experiments, and in endeavours to push his apparatus.

We now reach the superb serpentine cliffs of the Lizard with the beautiful coves of Polurrian, Mullion, and Kynance. At Lizard Point is one of the most famous of all lighthouses, the departure-point or landfall of thousands of ships in the course of the year. The peninsula of the Lizard is interesting, though the land does not rise much above 300 ft., and is monotonous moorland. All its charm is in its coast-line. The terrible Manacles rocks have been the scene of many a wreck. Helford river is a creek running up to Gweek in one arm and nearly to Constantine in another. We now reach Falmouth Bay, into which opens the Carrick Road. A curious peninsula, Roseland, runs to Zoze Point, where there is a lighthouse. Portscatho is a small place at the opening of Gerrans Bay, of which the eastern horn is

Mullion Cove

Nare Point. Carn Beacon is traditionally held to have been the burial-place of Geraint, King of Devon and Cornwall. There were more than one of this name. The cairn has been opened and was found to contain a stone cist of the bronze period, and not, as tradition said, his golden boat with silver oars. Veryan Bay between Nare Point and the Dodman has in it no good harbour. Dodman stands nearly 380 ft. above the sea.

Falmouth, from Flushing

Mevagissey Bay is a shallow hollow between Chapel Point and Black Head, the latter crowned by one of the cliff castles found on almost every headland. Then comes St Austell Bay with that of Tywardreath opening out of it. Charlestown has latterly become of importance, as from thence much china-clay is shipped.

We reach now the narrow estuary of the Fowey river, with Fowey town consisting of one narrow street beside the tidal creek, and with Polruan on the further shore. The coast now becomes very bold, and Polperro, five miles beyond, was once a notorious haunt of smugglers.

At Looe, the two rivers bearing the same name fall into a bay, and seaward stands up Looe Island, crowned

Polperro

by the ruins of a chapel. This island was also a haunt of smugglers, and it was found necessary to establish a coast-guard station on it to keep them in control. East Looe and West Looe each sent two members to Parliament before the passing of the Reform Bill.

Between Looe and Rame Head is Whitesand Bay, so called from the whiteness of the sand. The quicksands

Looe

have made it dangerous for bathers, but the cliff scenery is beautiful and romantic. There is a tiny watering-place at its western end, Downderry. At Tregantle is the most important of the western defences of Plymouth. A peninsula is formed by the Lynher river which discharges into the Hamoaze, and the neck of land between it and the sea is about two miles in breadth. Tregantle stands 400 ft. above the sea and commands every approach to Plymouth Sound.

Rame Head projects into the sea from Maker Heights and is the termination of a range of cliffs from Looe, and from hence a fine view can be had of the Cornish coast as far as the Lizard.

On the east of Penlee Point is Cawsand Bay, once infested with smugglers, sheltered by Rame Head from westerly gales. A rock with a cave in it and a white incrustation is regarded here with some superstitious reverence, and fishermen throw a few pilchards or herrings to it as an oblation when returning from fishing.

10. The Coast—Gains and Losses.

At a vastly remote period a valley lay between Britain and Gaul—before ever they were Britain and Gaul—and through this well-wooded valley flowed a river. The coast-line of Britain then lay from one to two hundred miles to the west, where is now the great drop in the ocean depths from 100 fathoms to 300 or 400. Cornwall was a Mesopotamia, a land between two almost parallel rivers, one occupying the bottom of what is now the English Channel, the other being the Severn. At that time the Bristol Channel was another great valley. From Brown Willy to the Scilly Isles ran a lofty mountain range, towering into the sky, of which the present Bodmin moors, Carn Brea, Carn Marth, Tregonning Hill, etc., are but the abraded stumps. Not only were they much higher, but their present roots stood 300 ft. higher than at present.

Then ensued a sinking of the land, and the Atlantic flowed into the valley to the south and joined the North Sea; and at the same time the Bristol Channel was formed. Thus the present coast was approximately outlined.

At some period shortly after, a vast inundation swept over the land from the north, and carried down the degraded granite, depositing the tin beds in the hollows. As early as 1830, Mr Carne noted: "The peculiar situation in which nearly all the stream tin of Cornwall is found is highly illustrative of the direction in which the current of the deluge swept over the surface. All the productive streams are in the valleys which open to the sea on the southern side of the Cornish peninsula; whilst most of the richest veins are situated near the northern coast."

The deposit of tin stone, or tin ground, lies directly on the *shelf*, or primitive surface of rock, and is carried far out in the estuaries, and overlaid by marine deposits.

In 1828, in Carnon Creek, a cairn was discovered 16 to 18 ft. below the surface, and that surface 4 to 5 ft. below low water mark. In it was a crouched skeleton. This shows that there must have been a subsidence of the land of something like 30 ft. at least since the period when man in the late stone or early bronze age inhabited Cornwall.

The submarine forests grew on the top of the tin ground. Of these many have been noted and recorded, not only on the south coast, but on the north as well. The trees were oak and hazel, alder and elm, but they never reached a large size.

Above this bed lie the raised beaches, some 40 or 50 ft. above high water mark. The tin-beds in the Cornish valleys towards the sea do not exhibit such an upheaval. Generally the raised beach rests on the original rock, and consists of rolled stones, frequently of large size, mixed with smaller gravel and sand. The " Head of Rubble," with some intervening perplexing beds of sand, may be noticed on the coast. This Head is from 40 to 50 ft. in depth. It is composed of angular fragments of rock, often large, many of quartz, with no signs of stratification. The Rubble bed in Cornwall has yielded no organic remains, but elsewhere in it have been found the bones of the mammoth, elephant, woolly rhinoceros, reindeer, etc. It was not formed by the disintegration of the subjacent rocks, but by aqueous transport. It owes its origin to a powerful force of water, acting violently and rapidly. It caps the heights, and is not in the valleys where the tin ground has been deposited. It has not therefore been found to overlie it. It was due to a sudden and brief overrush of water, and the fragments of stone carried before the flood did not travel sufficiently far to have their angles rubbed down.

As may well be supposed, the action of the sea on the coast-line has been affected largely by the character of the rocks against which the waves have lashed themselves to

Perranporth Rocks

foam. Where the rocks are of granite or slate, the tide and the waves have very little effect upon the outline of the coast; it is in those places where the softer rocks prevail, and where exposure to the prevalent wind induces breakers of great volume, that the loss of land by the action of the sea is greatest. In fact, the tides rarely run beyond one or two miles per hour, except round headlands, and it is where the rocks are of a yielding character that the coves and bays are formed. This is not always the case. Where the sea has found a fault in the rock it will burrow incessantly till it has bored out for itself a cavern, and the head of this falling in produces a tiny cove. The hard quartzose and trap rocks of Trevose Head, the greenstone rocks of Pentire, near Padstow, the hard slates of St Agnes, the greenstone and hardened schist of the Gurnard's Head, and the granite of the Land's End, defy the action of wave and tide. But it is otherwise where, for instance, the Head of Rubble occurs. "In Gerrans Bay it is plain that the cliffs of Head were at one time much further out than they are now. The tops of the earthy cliffs are split with cracks and miniature chasms, showing that great masses are constantly being detached by atmospheric causes, while the great heaps of earth at the foot of the cliffs show how for centuries masses of earth have rolled down from their tops on to the beach below[1]." As to the kingdom of Lyonesse, which was supposed to extend from the Land's End to beyond Scilly, it never existed in the historic or prehistoric period. Another fable that may be dismissed, is that St Michael's Mount was surrounded by a vast tract of woodland, in which were villages, and where settled hermits. It did rise

---

[1] D. C. Whitley, "The Head of Rubble," in *Journal of the R. Inst. of Cornwall*, XVII. p. 67.

above a forest, but that was in a prehistoric period. But if the sea gains on the land but imperceptibly in one way, it gains in another on the north coast, by the action of the wind carrying the sand inland, and overwhelming field after field. It is not a little curious to mark, nevertheless, how a small dribble of a stream will arrest the onward march of the sand-dunes.

At Constantine by Padstow as already said the old church is enveloped in sand-hills, so is that of St Enodoc. The Perran Sands have so encroached that they extend over a mile and a half inland and have in process of time swallowed up two churches and a village. The Gear Sands have even climbed a hill to the height of 300 ft. The Godrevy, Upton and Phillack towans have moved inland from St Ives' Bay and engulfed the residence of the ancient kings of Cornwall at Riviere.

The same phenomenon has not taken place in the south, but there the estuaries have been silted up by the wash from the stream tin works. Formerly boats could come up to Tregony. Now the Fal is choked with detritus for miles down. Restronguet creek bore vessels to Perranarworthal. Now it is completely silted up, only a trickle of water running down through desolate morasses and flats resulting from the workings of the miners.

11.   The Coast—Tides, Islands, and Lighthouses.

Off the mouth of the English Channel the tidal-stream is materially influenced by the indraft and outset of the channel, and is found to run northward and eastward with a falling tide at Dover, and southward and westward with a rising tide at that place.   At spring tides the tide rises in Padstow Bay 22 ft., at Bude a foot higher, at the Lizard only 14½ ft., at Scilly 16 ft.   Nowhere on the Cornish coast is there the enormous rise seen at St Malo, where ordinary tides rise from 23 to 26 ft., and spring tides 48 ft. above low-water mark.

On account of the varying force with which the channel and spring tides blend south of the Scilly group the stream is incessantly altering, but north of this towards the Bristol Channel, the stream becomes more regular, and while the water is ebbing at Dover, it sets northward turning sharply round Trevose Head into the Bristol Channel, and so when the tide is flowing at Dover, it is running with equal speed in ebb out of the channel and along the coast towards Scilly.

The Scilly Isles are the sole group of any importance around the coast.   They are situated 40 miles due west from the Lizard Point and 25 west-south-west from the Land's End, and are reached by steamers from Penzance. There are now but five of the isles inhabited, St Mary's, Tresco, St Agnes, St Martin's, and Bryher.   Formerly Sampson was also inhabited, but the inhabitants were

Round Island, Scilly

removed to St Mary's. The total acreage of the islands is 3560; and the formation is granite. This group is in fact the rubbed-down stump of the last great peak of the chain running south-west from Bodmin moors. The heights in the islands are inconsiderable, but very bold and picturesque scenery is obtained among the many islets, each of which has its special character. At St Mary's is

The Longships Lighthouse

a pier built in 1835–8; and a harbour called the Pool for small craft, while further out between the islands is a good roadstead for large vessels. The Scilly Isles were noted as a resting-place for innumerable birds, some very rare, in their annual migrations, but of late years gun-practice at sea marks has scared a good many away, and they visit the islands in far fewer numbers than formerly.

The coast of Cornwall is remarkably free of shoals. The only dangerous sandbank is the Doom Bar at the mouth of the Bay of Padstow; but there are shallows in Mounts Bay and other places.

Eddystone Lighthouse

Trinity House, the first general lighthouse and pilotage authority in the kingdom, was composed of a body of merchants and seamen founded in 1519 by Sir Thomas Speet, controller of the navy, when it was granted a charter by Henry VIII. Since that period the duty of erecting and maintaining lighthouses and other sea-marks has been entrusted to the Corporation by Royal Charter and Acts of Parliament. Trinity House maintains ten lighthouses about the coast of Cornwall, of which the most important, besides those of the Scilly Isles, are the Longships, white and red occulting; the Wolf rock, white and red group flash; the Lizard, white flash; the Eddystone, white flash.

12. Climate—Rainfall.

The climate of a country or district is, briefly, the average weather of that country or district, and it depends upon various factors, all mutually interacting—upon the latitude, the temperature, the direction and strength of the winds, the rainfall, the character of the soil, and the proximity of the district to the sea.

The differences in the climates of the world depend mainly upon latitude, but a scarcely less important factor is this proximity to the sea. Along any great climatic zone there will be found variations in proportion to this proximity, the extremes being "continental" climates in the centres of continents far from the oceans, and "insular" climates in small tracts surrounded by sea. Continental climates show great differences in seasonal temperatures, the winters tending to be unusually cold and the summers unusually warm, while the climate of insular tracts is characterised by equableness and also by greater dampness. Great Britain possesses, by reason of its position, a temperate insular climate, but its average annual temperature is much higher than could be expected from its latitude. The prevalent south-westerly winds cause a drift of the surface-waters of the Atlantic towards our shores, and this warm water current, which we know as the Gulf Stream, is the chief cause of the mildness of our winters.

Most of our weather comes to us from the Atlantic. It would be impossible here within the limits of a short chapter to discuss fully the causes which affect or control weather changes. It must suffice to say that the conditions are in the main either cyclonic or anticyclonic, which terms may be best explained, perhaps, by comparing the air currents to a stream of water. In a stream a chain of eddies may often be seen fringing the more steadily-moving central water. Regarding the general north-easterly moving air from the Atlantic as such a stream, a chain of eddies may be developed in a belt parallel with its general direction. This belt of eddies or cyclones, as they are termed, tends to shift its position, sometimes passing over our islands, sometimes to the north or south of them, and it is to this shifting that most of our weather changes are due. Cyclonic conditions are associated with a greater or less amount of atmospheric disturbance; anticyclonic with calms.

The prevalent Atlantic winds largely affect our island in another way, namely in its rainfall. The air, heavily laden with moisture from its passage over the ocean, meets with elevated land-tracts directly it reaches our

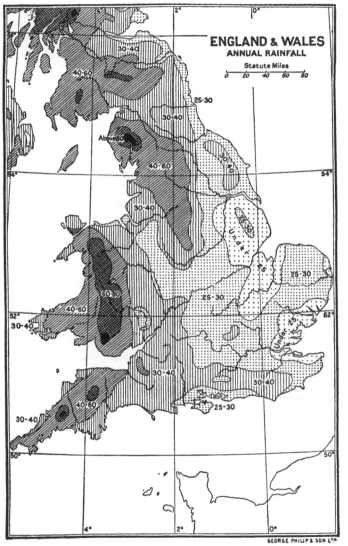

(The figures give the approximate annual rainfall in inches)

shores—the moorland of Devon and Cornwall, the Welsh mountains, or the fells of Cumberland and Westmorland —and blowing up the rising land-surface, parts with this moisture as rain. To how great an extent this occurs is best seen by reference to the accompanying map of the annual rainfall of England, where it will at once be noticed that the heaviest fall is in the west, and that it decreases with remarkable regularity until the least fall is reached on our eastern shores. Thus in 1906, the maximum rainfall for the year occurred at Glaslyn in the Snowdon district, where 205 inches of rain fell ; and the lowest was at Boyton in Suffolk, with a record of just under 20 inches. These western highlands, therefore, may not inaptly be compared to an umbrella, sheltering the country further eastward from the rain.

The above causes, then, are those mainly concerned in influencing the weather, but there are other and more local factors which often affect greatly the climate of a place, such, for example, as configuration, position, and soil. The shelter of a range of hills, a southern aspect, a sandy soil, will thus produce conditions which may differ greatly from those of a place—perhaps at no great distance—situated on a wind-swept northern slope with a cold clay soil.

The character of the climate of a country or district influences, as everyone knows, both the cultivation of the soil and the products which it yields, and thus indirectly as well as directly exercises a profound effect upon Man. The banana-nourished dweller in a tropical island who has but to tickle the earth with a hoe for it to laugh a harvest is of different fibre morally and physically from the inhabitant of northern climes who wins a scanty subsistence from the land at the expense of unremitting toil. These are extremes ; but even within the limits of a county, perhaps, similar if smaller differences may be noted, and the man of the plain or the valley is often distinct in type from his fellow of the hills.

Very minute records of the climate of our island are kept at numerous stations throughout the country, relating to the temperature, rainfall, force and direction of the wind, hours of sunshine, cloud conditions, and so forth, and are duly collected, tabulated, and averaged by the Meteorological Society. From these we are able to compare and contrast the climatic conditions in various parts.

Cornwall, being so surrounded by the sea, and so peculiarly under the influence of the Gulf Stream, may be looked upon as possessing a local "insular" climate in a marked degree. The south-west wind almost invariably brings rain and warmth together, for the high cold granitic moorlands are naturally calculated to arrest these warm airs and chill them, causing thereby a downfall of the suspended water. But Cornwall does not enjoy the amount of sun heat to ripen fruit that is obtained on the east coast of England, and it is exposed to furious gales from the west. "The gale from the west," says Polwhele, "is here no gentle zephyr; instead of wafting perfume on its wings, it often brings devastation." On the north coast every tree that is exposed to it is dwarfed and bowed like a curling wave, and the foliage in spring is

often cut and browned by the salt spray. Even tombstones
in the churchyards on the heights have to be backed up
with masonry, and the churches are low, as if cowering
from the blast. According to a Cornish proverb: "There
falls a shower on every week day, and there are two on a
Sunday." In Scilly, however, there is more sunshine and
less rain than on the mainland. The myrtle, geranium,
fuchsia, and hydrangea grow luxuriantly; the red geranium
at Penzance will cover the front of a house, and palms and
other exotics thrive there and at Falmouth. The fields of
narcissus and daffodils cultivated for the market would be
more beautiful if the blooms were not systematically picked
before fully open, to be sent to London. No gardens in
England exhibit such a wealth of exotics as those of Tresco,
Carclew, Enys and Penjerrick, and some others in the
district of the Fal estuary, which seems peculiarly favourable
for the growth of sub-tropical species.

In 1906, the mean temperature of England and Wales
was 49·3° Fahr., while that of Cornwall was 51·2.
The mean temperature of England in 1907 was 48·5, of
Cornwall 50·6. But there exists a considerable difference
between north and south. At Redruth it was in 1906,
50·1, whereas at Truro it was nine degrees higher.

The east wind prevails in October and is strong in
March, the south-east in June, the south-west is felt in
every month save April, but very little in December.
The west wind is most prevalent in August, least so in
May. The north wind predominates in December and
July.

The rainfall chart here given shows that Cornwall

lies for the most part in an area where from 40 to 60 inches are annually recorded, though a strip of the north coast from St Ives to Padstow and again from Boscastle to the northern limit of the county shows a fall of less than 40 inches. This is because the uplands have robbed the rainclouds of a considerable portion of their contents, and accordingly it is on these high moorlands that we find the greatest rainfall. Thus on the moors between Launceston and Bodmin from 60 to 80 inches fall, and even this latter figure is exceeded in the neighbourhood of Rough Tor and Brown Willy.

From observations taken at the Royal Institution of Cornwall at Truro, from 1850 to 1881 it appears that the rainiest months are November and December, and next to them January and July; April and May are the least rainy.

13. People—Race, Dialect, Population.

The original population of Cornwall would seem to have been what is now commonly called Ivernian, the same as Iberian, the underlying race everywhere in Western Europe from the western isles of Scotland to Gibraltar. When the Romans invaded and conquered Spain, they found there already in the east the Celts and in the west the Iberians, and they designated the more or less fused population, Celtiberians. So in Cornwall, there was this dark-haired, dusky-skinned race, and the Brythons, Celts, of the people of the Dumnonii. There were extensive settlements by Irish in the Land's

End district, the Lizard, and along the north coast, in 490—510, owing to the expulsion of the Ossorians and the Bairrche from their lands in Ireland. The Saxon also crossed the Tamar and peaceably settled in East Cornwall.

The language spoken was Brythonic, akin to, and originally identical with Welsh and that spoken in Lower Brittany. It was distinct in some points from the Goidelic of Ireland, the Isle of Man, and the Highlands of Scotland. The main difference was that the *C* in the latter became *P* in the former. Thus *Ken* or *Cen* in such names as Kenmare and Ciaran would in Cornish become Penmare and Piran.

In the reign of Edward I, Cornish was spoken in the south Hams of Devonshire, but in the sixteenth century it was dying out even in West Cornwall. Norden, writing in 1580, says: "Of late the Cornish men have much conformed themselves to the use of the English tongue, and their English is equal to the beste, especially in the eastern partes....In the weste parte of the countrye, in the hundreds of Penwith and Kirrier, the Cornishe tongue is most in use amongste the inhabitants." At Menheniot about 1540 the Creed and the Lord's Prayer were first taught the people in English. In 1678, at Landewednack, the last Cornish sermon on record was preached. In the reign of George III (1777) died Dolly Pentreath, the last to speak the language.

Two modern Cornish dialects exist. That in the east very naturally is assimilated to the Devonshire; but in the west it has formed itself in comparatively recent times. The more common prefixes of names of places are:—

Tre an enclosure or homestead, *Lan* also an enclosure for a church, *Ty* a dwelling, *Bod* a habitation, *Chy* a house, *Pol* a pool, *Pen* a head, *Huel* or *Wheal* a mine, *Ros* a moor, *Men* a stone, *Bāl* a mine, *Bos* is a corruption of *Bod* and is sometimes reduced to *Bo* ; *Car* stands for *Caer* a fortress; *Dun* has the same signification. *Burn* stands for *Bron* a hill, *Camborne* should be *Cambron*, the crooked hill; *Cara* is *Carreg* a rock, and this is often found cut down to *Car*; *Carn* is a cairn or heap of stones or mass of rock; *Enys* or *Innis* an island or peninsula. *Fenter* occurring in many combinations is *fenten* a well or spring ; *Goon* is the Welsh *Waun* a grassy down; *Hal* is a moor, *Parc* an enclosure, *Dinas* a chieftain's castle, *Lis* a court of justice.

It is possible that certain river names may derive from the original tongue of the earliest race. Some seem to be more akin to Goidelic than Brythonic dialect, as *Fal* (Gaelic *foill*, slow), and *Fowey* (Gaelic *fobhaidh*, swift); but there are not sufficient of these to assure us that the Goidels preceded the Brythons in Cornwall.

In East Cornwall there must have been a considerable influx of Saxon settlers, for there we light upon such names of places as are compounded with *ton* the home-farm, *worth* and *worthy* a fortified settlement, *stoke*, *stow*, *sto* a stockade, an earthwork surmounted by a defence of posts ; *ham* a meadow by the waterside. In West Cornwall such names are few.

The people are courteous and kindly, and very independent. In South Africa, whither many thousand miners have migrated, they are not popular, and are fain to

disguise whence they come by pretending to hail from
Furness.   The reason is that the Cornish cling together
and do not care to associate with others, and that they

Miners: Camborne

lack that breadth of sympathy which makes men give up
their time and devote their energies to the common good.
The Cornish miner cares nothing for the country and for
those among whom he has placed himself, if only in three

years he may have made enough money to return home to his wife and children.   He does not go out to settle, and this explains the lack of interest he has in what interests and concerns the colony.

The Cornish are a broad-shouldered race, above the average in stature, and it is stated that west country regiments, when drawn up on parade, cover a greater space of ground than would those of other counties, the numbers being equal.

The census of 1901 gives as the population of the administrative county 322,334 persons, of whom 149,937 were males and 172,397 were females.   There had been an increase in the number of males since 1891, and a decrease in the number of females.   Of inhabited houses there were 72,660.   The population of Cornwall in 1801 was 192,281, and it went on increasing steadily to 1861, but from that date it has been as steadily on the decrease. In military barracks there were 629 officers and men, in naval barracks 90, and on H.M. ships in home waters 2053; in workhouses 1308, in prison 75.

In 1901 there were 21,335 marriages, 88,636 births, and 55,790 deaths.   There were 6434 wives whose husbands were absent, either at sea or mining in South Africa or South America.   In agriculture 23,671 were engaged, in fishing 3734, in mining 13,426 ; in building and carpentering 9667.   Of the 149,937 males enumerated in the county of Cornwall, 128,184 were natives of the county, 8007 had come from Devonshire, and there were resident in Cornwall 976 foreigners.   Of blind there were 447, of deaf and dumb 163, of lunatics 798, of imbeciles 516.

Whereas the number of people to the square mile in England is 558, that in Cornwall is only 230.

It must, however, be born in mind that the Administrative County and the Geographical County are not coextensive. Thus, parts of the Registration County of Cornwall are in the Administrative County of Devon, such are Broadwood Widger, Northcott, North Petherwin, St Giles in the Heath, Virginstow and Werrington, with a total population of 2460 persons; and on the other hand parts of the Administrative County of Cornwall are in the Registration County of Devon; these are Calstock and North Tamerton with a population of 6203 persons.

Thus the population and acreage in 1901 in the Registration County would be, population 318,591, acreage 886,384, but in the Geographical or Administrative County the population would be in the same year 322,334 and the acreage 868,208.

14. Agriculture — Main Cultivations. Stock.

The lack of hot sunshine makes Cornwall an unsuitable county for cereals, but the mildness of the climate and the rainfall render it on the other hand favourable for dairy produce and for stock.

A recent writer, Mr W. H. Hudson, in his book on the Land's End, thus describes a Cornish farm. He speaks of the neighbourhood of Penrith, but the description applies to many other parts of the Duchy, though not to East Cornwall, where it is scarcely applicable at all, and is

precisely the part where there exists least of the Celtic and
most of the English element.   "Life on these small farms
is incredibly rough.   One may guess what it is like from
the outward aspect of such places.   Each, it is true, has its
own individual character, but they are all pretty much
alike in their dreary, naked, and almost squalid appearance.
Each, too, has its own ancient Cornish name, some of them

"First and Last House," Land's End

very fine and pretty, but you are tempted to rename them
in your own mind Desolation Farm, Dreary Farm, Stony
Farm, etc.   The farmhouse is a small, low place, and
invariably built of granite, with no garden, or bush, or
flower about it.   The one I stayed at was a couple of
centuries old, but no one had ever thought of growing
anything, even a marigold, to soften its bare, harsh aspect.

The house itself could hardly be distinguished from the outhouses clustered around it. Several times on coming back to the house in a hurry, and not exercising proper care, I found I had made for a wrong door, and got into the cow-house or pig-house, or a shed of some sort, instead of into the human habitation. The pigs and fowls did not come in, but were otherwise free to go where they liked. The rooms were very low ; my hair, when I stood erect, just brushed the beams ; but the living room or kitchen was spacious for so small a house, and had the wide old open fireplace still so common in this part of the country. Any other form of fireplace would not be suitable where the fuel consists of furze and turf."

But it must be allowed that the large landed proprietors have everywhere built commodious, though almost invariably ugly, new farmhouses.

Considering the vast extent of grass land there is in Cornwall, and the amount of butter, cream, and milk derived therefrom, it has become more and more clear to thinking farmers that the pasture is becoming exhausted, its feeding powers worn out, and that they must replenish the soil with the lime and phosphates that have been taken out of it to nourish and rear the cattle. The consequence is, that of late years the more intelligent farmers are largely employing artificial, i.e. chemical manures, and the results have been most satisfactory. You cannot eat your cake and have it. If you take so much annually out of the ground you must put in an equivalent annually, or you exhaust the soil ; and animal manure is not sufficient. In the country a great deal of turf and furze is burnt, where there is proximity to the moors. The writer just quoted says : " in some parts of Cornwall they have good peat, called 'pudding turves,' which makes a hot and comparatively lasting fire. In the Land's End district they have only the turf taken from the surface, which makes the poorest of all fires, but it has to serve. But to make a blaze and get any warmth furze was burnt. In a few moments the dry stuff would ignite and burn with a tremendous hissing and crackling, the flames springing up to a height of seven or eight feet in the vast hollow chimney. For a minute or two the whole big room would be almost too hot, and lit up as by a flash of lightning. Then the roaring flames would sink and vanish, leaving nothing but a bed of grey ashes, jewelled with innumerable crimson and yellow sparks, rapidly diminishing."

The total acreage under crops and grass in 1908 was 608,691; of these 356,497 acres were arable land, the rest, 252,194 acres, permanent grass : 17,120 acres grew wheat, 30,696 barley, and 66,033 oats; 4441 acres grew potatoes, 15,271 turnips and swedes, 11,528 mangold, 8059 rape, and 1658 small fruit. Of small fruit culture just over 621 acres were devoted to strawberries, nearly 276 to raspberries and nearly 553 to currants and gooseberries. Apples were grown on 4865 acres, cherries on 199, plums on 182.

The number of horses used for agricultural purposes was 25,706. The total of horses, colts and unbroken, was 34,821. Of cattle there were 219,890, and of these 59,298 were cows in milk. The total number of sheep was 410,055, and that of pigs 104,813. Of mountain and heath land used for grazing there were 71,438 acres. In Cornwall there are 9249 acres of coppice and 22,197 of other woods, but no more than 981 of plantations. While 548,787 acres are tended by tenants, only 59,904 acres are occupied by the owners. Whereas in Surrey the percentage of those occupying their own land is 33 per cent., that in Cornwall is 9·7, in Devon 10·9, and in Hampshire 24·4.

In Cornwall, in 1908, the number of agricultural holders above one acre but not exceeding five was 2859; of those above five and not exceeding 50, there were 6810. Of those above 50 and not exceeding 300 acres there were 3682, but of those holding more than 300 acres there were but 118, whereas in Essex there were 564, in Suffolk 540, in Hampshire 595, in Wiltshire 660 and in Northumberland 731.

Of late years, especially in the Scilly Isles, flower gardening, the growth of narcissus, jonquils, daffodils, and various sorts of lilies, also of violets, anemones, and marguerites, has been carried on with great success, and the produce is carried by steamer and train to London and other large towns.

During six months in the year, when and where the flower culture prevails, it forms the staple of conversation in parsonage, manor house, farm, and cottage. From January until May men, women, and children are directly or indirectly engrossed in this one labour. The bulbs are moved at least once in three or four years. Left longer, they decrease in size, and become weakly;

the flowers also degenerate. A suitable manure is at hand, the kelp washed up by the sea costing nothing but the labour of gathering and transporting.

Flower Farming, Scilly

Anyone standing in a field which is in full bloom and waiting to be picked over, would think that the men have a hopeless task to face, if they purpose gathering all their flowers. But they move down the beds swiftly, snapping the stems, and throwing the flowers into big baskets, which are carried off to the homestead as they fill, and in an incredibly short time the beds are thinned. When the baskets are brought into the farmhouse they are emptied, and if the weather has been wet and stormy the flowers are packed roughly into pots and pans of every description, and set on shelves to dry off and assume their proper colour. If the weather has been fine this preliminary toilet is dispensed with, and the girls and women bunch and tie at once. Twelve stems go to a bunch always, and the aim is to arrange the flowers so that they shall present a compact lozenge shape, crisp and tight. Varieties are never mixed when tied, the bunches are passed on to another department, where the uneven stems are sliced off, and the flowers set in water to await packing.

Packing the flowers is a serious business. The boxes are lined with paper, little pillows are made for the flowers to rest upon, and then the bunches are deftly laid, so many this way, so many that, and a few in the centre, and behold, stems are altogether hidden, and only a mass of bloom fills the box. The end papers are turned in, a ticket placed on the top, declaring the variety of flower and the number of bunches; the cover is nailed down, and the operation is complete.

15. Industries and Manufactures.

Cornwall is too far from the coalfields to be a manu-
facturing county on a large scale.   There are, however,
some few industries and manufactures carried on in
Cornwall.   The wonderful wireless telegraphy installation
which Mr Marconi has established at Poldhu does not,

Poldhu Hotel and the Marconi Station

perhaps, strictly speaking, come under this head, but it
would be impossible to omit mention of it and of the
great services it daily renders to vessels crossing the
Atlantic.

A considerable industry is in the making of casks,
mainly for the exportation of pilchards.   Woollen manu-

facture is carried on in a small way, also the construction of mining and other machinery. Shipbuilding occupies about 870 men. Some brick-making is done, but not to any considerable extent. The real employments supplying the vast majority of the people with bread are mining and quarrying, and agriculture and horticulture.

16.   Minerals and Mining.

The authentic history of tin-mining in Cornwall begins with the year 1156. It is not mentioned in Domesday, and probably the Conqueror was as ignorant that tin was to be found there as were the Romans. In the year above given the tin mines are mentioned in the Pipe Rolls. In 1198 appears a letter from the Warden of the Stannaries to the Justiciar. In 1156 most of the tin was raised on Dartmoor, but the output now began to rise rapidly. In 1201 King John issued a charter to the Stannaries. In 1305 the Cornish Stannaries were in part dissociated from those in Devon.

As already stated, in the beds of the valleys running down from the central ridge are deposits of tin, brought there from the lodes degraded by weather and flood in the central spinal ridge. The specific gravity of the tin ore is 6·8, and as the water rolled away from the heights, it deposited the tin that it had brought away with it. "The ores of tin," wrote Pryce in 1776, "are shode and stream...scattered to some distance from the parent lode, and consisting of pebbly and smoothly angular stones of various sizes, from half an ounce to some pounds in weight. Stream tin is the same as shode but smaller in size and arenacious."

Polwhele describes an early stream work disclosed about a century ago. The ore was of the purest kind and contained two-thirds metal. The pebbles from which the metal was extracted were in size from sand-like grains to

King Edward Mine, Camborne

that of a small egg. The depth of the primeval bed was 20 ft. This appeared to have been worked at a remote period, and before iron tools were employed, as large pick-axes of oak, holm, and box were found there. But archaeological research at the period when he wrote was carried on in such a haphazard way that we cannot trust the reports then made, and it is quite possible that the workings were comparatively late, possibly of Mediaeval times.

G. C.   6

All the early work was in stream tin. The Cornish tin had this disadvantage—that it was not pure like the Dartmoor tin, but was so associated with sulphur that after smelting it had to undergo a second process, and that a delicate one. It had to be "roasted," to get rid of the sulphur, for only thus could the tin be made available. Associated with sulphur it is brittle. We may well doubt whether in early times the double process was understood. Moreover, we know that when we have the first notices of tin in the west, it was Dartmoor and not Cornwall which rendered the largest supply.

Shaft-mining did not come in before 1450, when the stream tin was exhausted. The use of the adit cannot be traced back beyond the beginning of the seventeenth century. The only way of draining the mines was by rag and chain pumps, each consisting of an endless chain, broadened at intervals by leathern bindings, to fit snugly into a pipe from 12 to 22 ft. long, worked by a windlass at the surface. To drain a mine of great depth a series of these pumps was necessary.

When hydraulic draining-engines were first employed is not known, but even so late as the close of the eighteenth century some mines were drained by the rag and chain pump worked by 36 men.

Up to the sixteenth century, wooden shovels and picks are known to have been employed, and shovels were merely iron-shod. There is in the British Museum a MS. calendar of Haroldstone in S. Wales, of the early sixteenth century, in which are representations of the works of the months, and in it the labourers in the fields are shown as using wooden spades shod with metal.

German mining was carried on upon better principles than the English, and Sir Francis Godolphin sent for an experienced German engineer to instruct the miners of Godolphin and Tregonning in the superior systems employed in Saxony. It was then only that the hydraulic stamps were introduced.

In 1742 one steam-engine only was found in the county; but speedily after came a great advance. Savery and Newcomen brought in their steam pump ; that of Newcomen was introduced at Chasewater in 1777.

Notwithstanding that the methods of descent into the mines by a series of long ladders had been superseded by the man-engine, first introduced in 1842, it was a long time before the old ladders at different successive stages were abandoned.

Owing to the introduction of tin from the Straits Settlements, where it is found in the condition of stream tin and can be easily worked, the tin mining in Cornwall, necessitating real mining and the following of lodes, has proved unremunerative and has been abandoned, and now but few mines are worked for the metal. Another cause has fatally affected Cornish mining—the fraudulent practice of the promoters of the mines. It was no uncommon practice for the "captains," when a rich lode was struck, to cover it up, and follow false lodes, till the investors in the venture lost heart and refused to advance more money, when the captains would carry on the work in the real lode, if they could raise the capital, but this they often failed to do,

6—2

the mine having fallen into discredit, or the water having broken in.

But if tin mining be practically dead in Cornwall, another industry has risen with leaps and bounds. It is that of the china-clay and china-stone, employed in the manufacture of porcelain, in the sizing of paper and of cotton materials, in the manufacture of alum, etc. The

China Clay Quarries, Porthpean

glazed paper so largely employed in our illustrated papers is made up largely with china-clay. Some years ago the Italian government employed this paper for its official documents, but found that after a few years under the influence of damp weather the records had dissolved into a lump of clay.

China-clay consists of decomposed felspar, quartz and

white mica. In 1817 the amount shipped for manu-
facturing pottery was comparatively small, but of late
years it has grown, and employs over 3000 persons. The
china-clay is the same as the Chinese *Kaolin*; its value
was discovered by a Plymouth Quaker, Mr Cookworthy,
in 1745.

Porcelain was introduced into Europe from the East
in 1518, when it acquired the name of China. For a
long time it was supposed that the kaolin or fine white
clay of which it is composed, was found only in the
Celestial Empire, and specimens of this brought to Europe
fetched a high price. At the beginning of the eighteenth
century it was discovered in Saxony in an odd way. A
merchant named Schnerr, being on a journey, was struck
with the whiteness of some clay near Schneeburg, and
collecting some of it used it for powdering his wig. It
succeeded, but had this disadvantage, that wigs dressed
with this new powder were very heavy. An apothecary
named Bötcher noticed the increased weight of the wig,
analysed the powder, and discovered that it was identical
with Chinese kaolin. He began to make Dresden China
in 1709, and the process was carried on with the greatest
secrecy, the exportation of the earth being forbidden under
heavy penalties.

In 1748 Cookworthy discovered kaolin on Tregonning
Hill, more was found at Boconnoc, and Cookworthy and
Thomas Pett began to make china in 1768. At present
St Austell is the great seat of this industry, and the produce
is shipped at Charlestown and Polmear.

The important slate-quarries of the Duchy, of which

Delabole Slate Quarries

the Delabole quarry is the most renowned, have already been mentioned.

At Calstock and Gunnislake a few years ago there were numerous miners engaged in the manufacture of arsenic from the waste product of the abandoned copper mines. But now this has become an extinct industry.

Shipping Slate, Port Gavin

Wulfram or tungsten, a metal used as an alloy for hardening steel, was also a waste product from the tin mines, but it is now utilised. At St Ives, pitchblende is now being worked for radium.

17. Fisheries.

If mining be a decayed industry in Cornwall, that of fishing shows no diminution.   In an old book of natural history published in 1776, the principal fishery of pilchards is described.   "Pilchards appear in vast shoals off the Cornish coasts about the middle of July, and disappear at the beginning of winter; though a few of them sometimes return again after Christmas.   The fishing employs a great many men on the sea, and men, women, and children on land, in salting, pressing, washing, and cleaning; in making boats, nets, ropes, casks; and all the tradesmen depending on their construction and sale.   The usual quantities exported each year, for ten years, from 1746 to 1756 inclusive, on the average is as follows: Fowey has exported 1732 hogsheads annually; Falmouth 14,631; Penzance and Mounts Bay 12,149; St Ives 1282; in all amounting to 29,795 hogsheads."   And the same writer thus describes the fish.   "The pilchard greatly resembles the herring, but differs from it in some particulars; it is a third less, and the body is proportionably broader: it has a black spot near the upper corner of the gills, and the belly is not so sharp.   It has no teeth, either in the jaws, the tongue, or the palate."   It is now held that the pilchard is identical with the sardine, but in a different stage of growth.

The pilchards are taken generally from the middle of August to the middle of September, when large "schools" are seen coming up the Channel.   Each fishing station generally has two or more companies or clubs of twenty or thirty men; each company owning various boats and generally two of the gigantic seines employed, which cost £250 or more. These nets are about 250 fathoms or more long and about 15 fathoms deep, and three boats go to each seine. The first boat, which is also the largest, is called the seine-boat, as it carries the net and seven men

Pilchard Boats, Mevagissey

in it; the next is termed the "vollier," or "cock-boat," and carries another seine, called the tuck-seine, which is 100 fathoms long and 18 deep, this boat also carries seven men; the third boat is called the "lurker," and contains three or four men, and in this boat is the master.

The pilchards were at one time supposed to come from the Polar Sea, but it has now been ascertained that the main body retires for the winter into deep water to the westward of the Scilly Isles. About the middle of spring they quit the deep seas and begin to consort in small shoals which gradually increase to the end of July or the beginning of August, when they combine in one mighty host and begin their migration eastward. They strike the land a little north of Cape Cornwall, where they break up into two portions, one following the north coast of Cornwall, the other the south.

When the shoal is approaching, men and boys who have been lying on the cliffs doing nothing start into activity and rush to the boats. The gulls may be seen hovering over the advancing army, and a change appears in the colour of the water. At once the "huers," as the signallers are called, get out their signals—a ball at the end of a stick—and proceed to direct the pursuing boats according to the movements of the school.

The seine-boat leads the way followed by the vollier, and the crew of the foremost boat pass a warp, that is, throw a rope, which is fixed to the end of the seine on board the vollier, and then shoot the net overboard, which, having leaden weights at bottom, sinks, while the top is buoyed up with corks. The seine-boat is rowed in a circular course round where the fish are "stoiting" or jumping, and when they have reached the vollier, the fish are enclosed. They then join the two ends of the seine together with a cord to prevent the fish from breaking out, and whilst this is being done a man is engaged in frightening the fish away from the still open end by means of a stone fastened to a rope. When the two ends of the net are laced together, grapes, i.e. grapples, are let down to keep the net expanded and steady till the fish have been taken up. This latter process is called "tucking the seine." The boat with the tuck-seine on board passes the warp of that seine to one of the other boats and then shoots this net within the big seine. It is then drawn up to the edge of the water, when it is seen to be one quivering

The Huers' House, Newquay

mass of silver. The fish are now taken or dipped out with baskets into the boats. When the boats are filled, if more fish remain in the large seine, it is left in the water, till by successive tuckings all the fish have been removed. In addition to seining large numbers of pilchards are taken in drift nets.

Formerly pilchards were smoked, and went by the

name of *fumados*, which name has been corrupted into
"fair maids." A not over-complimentary saying in Cornwall is that "pilchards and women when they are bad are
very bad, and when good are only middling." Pilchards
constitute an important article of food to the poorer classes
on the coast, but doubtless the Cornishmen get very tired
of them as an article of diet. Large quantities are exported
to Spain and Italy. Some are made into "sardines" in oil
in the local factories.

A peculiarity of the county is that ecclesiastical dues
in the nature of tithes are levied on the persons employed
in the pilchard fisheries.

The number of hogsheads packed for export every
year varies considerably, but the yearly produce averages
from 20,000 to 30,000 hogsheads. In certain years the
amount has reached nearly 50,000. In 1901 the fishery
found employment for 3734 men, in 1905 in Penzance
alone for 1275. Beside pilchards, there are mackerel,
hake, cod, etc.

The quantity of fish taken, other than mackerel,
herrings, pilchards, and sprats in 1905 at Looe was 5841
cwt., at Mevagissey 4893, at Falmouth 3213, at Porthleven 6132, at Newlyn 37,468. Of mackerel, herrings
and sprats, at Looe 10,403 cwt., at Mevagissey 40,236, at
Falmouth 5991, at Porthleven 26,945, at Newlyn 493,956.
The total value of the fish taken in the year at Looe was
£5377, at Mevagissey £13,818, at Falmouth £17,718, at
Porthleven £11,454, at Newlyn £232,466.

Turning to the north coast, in the first category were
taken at Sennen 1074 cwt., at St Ives 2431, at Padstow

Pilchard Boats, St Ives Harbour

4051. In the second category, at Sennen 605 cwt., at St Ives 80,557, at Padstow but 318, at Port Isaac 2526.

The value of the fish taken at Sennen was £2065, at St Ives £39,941, at Padstow £6660, at Port Isaac £2169.

Landing Fish, Newlyn

18. Shipping and Trade.

There are no great seaports in Cornwall. Falmouth was by far the most considerable when the packet service ran from it to the West Indies, Portugal, and New York. The station was established in 1688. In 1705 five packet-boats left it for the West Indies and in 1709 as many for Lisbon; not till 1755 did two sail for New York, but the number was increased to four in 1763.

In 1782, eighteen packets sailed from Falmouth for the West Indies and America.   Up to 1823 the packets had sailed under contract between the General Post Office and the commanders, who received their appointment from that establishment and engaged to provide, equip, and man a proper ship for the purpose, for a sum of £1800 per annum.   These vessels were from 180 to

The St Vincent's Anchor, Falmouth

210 tons register.   But after 1823 the above system was changed, for the service was placed under the orders of the Admiralty instead of the Post Office ; and as vessels were wanted they were supplied by men-of-war.

In 1827 thirty-nine packets left Falmouth ; in 1834 six steamers were employed.   But in 1850 Falmouth ceased altogether to be a packet station.   This was greatly to the detriment of the town. It still remains as a port of call for outward-bound sailing-ships. Further up the river is Penryn, which was a town and a port before Falmouth was thought of. The silting up of the river does not now allow other than small boats and barges to reach it.

Letters came down on mail coaches, from London through Exeter, by Launceston to Bodmin, and thence to Truro and Falmouth. The rate appointed for the coaches, including stoppages, was bound to be thirteen miles an hour. The mail spun along night and day, without a halt save for change of horses. The stages on an average were eight miles, and the horses, four-in-hand, went at a gallop. The guard wore the royal livery of scarlet, and always had his blunderbus handy, in case of an attempt by highwaymen to hold up his coach.

Charlestown and Polmear in St Austell Bay are only important for the trade in china-clay exported thence. Marazion and Penzance, and St Mary's in Scilly only for the conveyance of flowers to London. But these ports and such as are on the north coast are convenient as mouths through which Welsh coal can be imported to feed the cellars and fires in the peninsula. Pilchards also are exported from these little ports to Italy and Spain; and anciently a considerable trade was carried on between them and France, Spain, and Portugal in wine, and a considerable amount of wine and spirits entered the county through small creeks and coves, into which smugglers conveyed their kegs. The gentry and taverners were kept well supplied with liquor that never paid duty.

19. History.

The original population of Cornwall was probably Iberic, of the same primitive race as the dark-haired population of Ireland, before the island was invaded and subjugated by the Celts.

The branch of the Celts in Britain and Cornwall was Brythonic, and there is nothing certain to show that the Goidels were in Cornwall before the Brythons. It is true that some few river names, and again inscriptions are Irish, but these latter pertain to the settlement in Cornwall of Irish expelled from Ossory and Wicklow in the fifth century.

Popularly the dark hair and dusky complexions of some of the Cornish is attributed to Spaniards wrecked from the Armada. But no Spanish wreck came on shore in Cornwall. The first loss the Armada sustained was east of Plymouth. On its way back to Spain none of the vessels came near Cornwall. Several were wrecked on the coast of Ireland and their crews massacred to a man by the natives.

Posidonius travelled after B.C. 123 and visited Spain, where he collected a variety of information on points of geography and natural history, and after spending thirty days at Gades returned to Italy. He learned among other things something about the collection of tin in Bolerium, a name afterwards appropriated by Ptolemy to the Land's End. He says: "The inhabitants of that promontory of Britain which is called Bolerium are very fond of strangers, and from their intercourse with foreign merchants are civilised in their manner of life. They prepare the tin, working very skilfully the earth in which it is produced. The ground is rocky, but it contains earthy veins, the produce of which is ground down, smelted, and purified. They make the metal up into slabs shaped like knuckle-bones, and carry it to a certain island lying in front of Britain called Ictis. During the ebb of the tide the intervening space is left dry, and to this place they carry over abundance of tin in their waggons....Here then, the merchants buy the tin from the natives, and carry it over to Gaul; and after travelling overland for about thirty days, they finally bring their loads on pack-horses to the outlet of the Rhone." It is very doubtful whether Posidonius ever visited Britain. What he relates is doubt-less due to information received by him, either at Gades or at Massilia.

According to Timaeus, the contemporary of Pytheas, the isle of Vectis was six days sail from Britain, "in an inward direction." Vectis, there can be little doubt, is the Isle of Wight, formerly connected with the mainland by a ridge of chalk since broken through by the waves. Ancient mariners coasted, and those who came to Britain for tin followed the Gallic shore till they could see the white cliffs of Dover, when they crossed, and coasted down channel to the Isle of Wight.

There is no evidence that the Phoenicians ever visited Cornwall. Nor has a single relic of Phoenician art or coin been found in Cornwall. The traders with Britain were the Veneti of the Morbihan, in Brittany. Moreover, as General Pitt Rivers has pointed out, bronze celts (axe heads), which have been unearthed in Cornwall, are never found in any parts where the Phoenicians have been.

It has been assumed with much confidence that Cornwall or the Scilly Isles must have been the Cassiterides of the ancients. But even this is doubtful. The Cassiterides were described as lying west of Spain, and the description

The Nine Maidens, St Columb Major

applies to the Azores; it may have been due to ignorance or design that they were· represented as islands prolific in tin.

That tin was worked in Cornwall from a very early period can hardly be questioned, in 1823 at Carnon a deer-horn pick was discovered 40 ft. below the surface, but as a crucifix was also found there 30 ft. below the surface, this only shows how the creeks have had their floors turned over and silted up.

Though tin was exported from Cornwall, bronze was not manufactured there till a comparatively late period. Bronze came from the East, and the great centre whence radiated the trade in bronze weapons was the basin of the Po.

What seems to be abundantly clear is that the export of tin from Britain had come to an end by the first century of the present era. Caesar, on invading Britain, heard nothing about it, and when Britain was finally conquered, the Romans who worked the lead mines in the Mendips, and gold and copper in Wales, totally neglected Cornwall, holding it to be worthless. They never settled there, only traversed it to the Land's End, leaving behind them a couple of square camps, some coins, few and far between, and some Samian ware; but this shows little more than that a traffic went on between the Britons of Cornwall and the Romans and Romano-British beyond Exeter. The fact that the Romans had no idea that tin was to be found in the peninsula shows that the mining for it had ceased there for some time previous. The Brythonic Celts are held to have invaded Britain B.C. 300; and it is probable that from that date the industry in tin mining carried on by the Ivernian natives declined rapidly and expired, leaving not a tradition behind.

It is noteworthy that the Dumnonii were behind the British peoples in the east. They had no coinage, whereas those in the south-east had theirs, copied from a stater of Philip of Macedon. No Greek coins had reached the

Dumnonii, and they had consequently none to copy, however clumsily.

In the battle of Deorham, 577, the Britons were defeated with great slaughter, and the West Welsh of Devon and Cornwall were cut off from further communication with their brethren of Wales. The Saxons steadily advanced, but for long the Parrett was the boundary. In 823 a battle was fought between the Saxons and Britons at Gavulford, now Galford, a point on the old road from Exeter to the west, where the hills draw together, and whence it is commanded by a huge camp. The Britons now called the Danes to their aid, and twelve years afterwards a battle was fought on Hingeston Down, above Calstock, in which Egbert was victorious. This was in 835. Hitherto the Britons had occupied one portion of Exeter, but Athelstan, after defeating the Cornish King Howell, not only expelled them from the city but fixed the Tamar as their boundary. Then he passed through Cornwall, and even visited the Scilly Isles. The Count of Poher in Brittany, of whose son Alan Barbetorte was the godson of Athelstan, fled from Brittany with a crowd of his countrymen from the devastations of the Northmen, and Athelstan gave them homes in England. It is probable that he planted some in the Lizard, and others about Camborne, for we find there church dedications to distinctively Breton saints, and we know, moreover, that the fugitive Bretons brought with them the bones of their patron saints. As they spoke the same tongue as the Cornish, it would be natural for Athelstan to send them there.

In 981, the Danish pirates plundered the monastery of St Petrock, and in 997 ravaged the territories of their old allies from one end of Cornwall to the other with fire and sword. Shortly before this, in 993, Olaf Tryggvason of Norway, with Sweyn Forkbeard of Denmark, after having attacked London, devastated the east coast, burnt Sandwich and Ipswich, and stormed Bamborough, then harried the Scottish coast, the Western Isles, the Isle of Man, and Ireland, where "he burned far and wide, wherever inhabited." He then came to the Scilly Isles, where he put into Tresco harbour. There the monks of the abbey founded by Athelstan so impressed him, that he consented to be baptized.

In 1068, the county was plundered by Godwin and Edmund, sons of Harold, after a battle in Somersetshire, and on their way back to Ireland.

In 1322, the craze for going on pilgrimage took possession of Cornish men, women, and children, and they set off for the Holy Land, whence few returned.

In 1497, in consequence of the dissatisfaction occasioned by the levy of a burdensome tax, the commoners of Cornwall, headed by Thomas Flamank, a gentleman, and Michael Joseph, a Bodmin blacksmith, rose in rebellion. Having prevailed on Lord Audley to be their general, they marched as far as to Blackheath in Kent, where they were defeated with much slaughter by Lord Daubeny. In the same year Perkin Warbeck landed, according to some at Whitesand Bay, near the Land's End, according to others in St Ives' Bay, and marching to Bodmin, found the Cornish ripe for a new rebellion. At the head of

3000 men he marched to Exeter, but was unable to take it. He made his way to Taunton, where, despairing of success, he deserted his army and fled, but was taken and executed.

In 1548, another rebellion broke out occasioned by the changes in religion, which the Cornish resented. Their leader was Hugh Arundell, Governor of St Michael's Mount. They proceeded to besiege Exeter, but although the city was sorely distressed for want of provisions it held out till relieved, and in a battle fought at Woodbury, they were defeated with immense slaughter. Cornwall remained quiet in the reign of Elizabeth, save that it sent out whole fleets of privateers to prey on the Spanish traders and treasure vessels.

The Armada was off the Lizard on the 29th to 30th July, 1588. An English fishing-boat was hanging near them, counting their numbers. They gave chase, but the boat shot away down wind and disappeared. It carried the news to Drake at Plymouth, who at once prepared to sail forth.

In the Civil War, Cornwall was almost unanimous on the side of the King. Lord Robartes, however, threw in his lot with the Parliament. On the side of the King were Sir Bevil Grenville, Sir Ralph Hopton, Godolphin, Slanning, and Trevanion. A battle was fought at Boconnoc on January 19, 1643, in which General Ruthven and the Roundheads were defeated. Another battle in which the Royalists were successful was at Stamford Hill above Stratton, May 16, 1643.

Charles I visited Cornwall, and was so impressed by

the devotion and loyalty of the people that he addressed
to them a letter of recognition, copies of which may be
seen in some of the churches.   Prince Charles spent a
great part of the autumn and winter of 1645 in Cornwall;
on March 2, 1645–6, he embarked at Pendennis Castle
for the Scilly Isles, where he "was much straitened for
provisions."   He quitted Scilly on April 16, and landed
next day in Jersey, whence he sailed for France.   Queen
Henrietta Maria had left Pendennis for France in July,
1644.

Cornwall took no active part in the Revolution; in
the European War, it sent forth many gallant sailors,
among whom in the first place may be reckoned Admiral
Boscawen, "old Dreadnought."   But since the Civil War
the history of the Duchy has been mainly one of social
and industrial advance.   The principal events stirring the
community were the introduction of steam-engines to
pump the mines, affrays with smugglers, and the excite-
ment and unlimited bribery and corruption at elections in
the rotten boroughs till these latter were swept away by
the Reform Bill of 1832.   About these rotten boroughs
a few words must be said.   The old boroughs that existed
before the reign of Edward VI were Truro, Helston,
Lostwithiel, Bodmin, Liskeard, and Launceston.   But the
advisers of Edward VI, conscious of insecure tenure of the
throne and doubting whether the country was willing to
go with them in their sweeping alterations in religion, and
desirous of counteracting the growing importance of the
House of Commons, considered that their object would be
best attained by conferring the right of returning members

of Parliament upon the obscure dependent villages of Cornwall. Accordingly Saltash, Camelford, West Looe, Bossiney, Grampound, Penryn, Mitchell, and Newport were elevated into boroughs, each returning two members to Parliament.

Under Queen Mary, St Ives received the same privileges, and under Elizabeth six more were made boroughs

St Mawes, Falmouth Harbour

with the same rights, St Germans, St Mawes, Tregony, East Looe, Fowey, and Callington. Some of these places, as Mitchell, Tregony, and St Mawes, were mere hamlets. They all soon passed away from the direct control of the Crown and fell into the hands of borough mongers who returned what members they liked, by gross bribery, expecting to be repaid with Baronetcies and with lucrative sinecures by the Ministry of the day they supported.

20. Antiquities : Prehistoric, Roman, Celtic, Saxon.

The first men who inhabited our island were the merest savages. They had no knowledge of the use of metals, they could not make pottery ; they had not domesticated the cow, the sheep, or the dog. They used extremely rude flint weapons and tools. They were contemporary with the cave bear, the woolly rhinoceros, the mammoth, and the cave hyæna, all which beasts then lived in Britain ; and at that time the temperature was much colder than at present. This period is called the Palaeolithic, or Old Stone Age.

The next race that entered our island found the temperature much as it is now. They were comparatively civilised. They still used flint implements, but of a very superior type, and far better finished than those of the earlier race. Moreover they were agricultural, grew corn, had cows and sheep and dogs, and made pottery. This race it was which erected the so-called cromlechs, stone circles, and tall upright stones. The remains of their villages of circular stone huts are very numerous on the moors. This period is called the Neolithic, or New Stone Age.

After a time bronze was introduced, by trade, and was at first as valuable as gold is to us. But after a while it became much more common. Its introduction marked the commencement of the Bronze Age.

To the Bronze Age succeeded the Iron Age. This

metal was introduced by the conquering Celts, of whom there existed two branches, one called Brythonic, of which were the Britons, now represented by the Welsh, the Cornish, the Cumbrians, and the Bretons of Brittany ; the other called Goidelic, and now represented by the Irish, the natives of the Isle of Man, and the Scots. The Celts with their iron weapons speedily overcame and enslaved the earlier race, usually called Ivernian. The latter was dusky and dark haired, but the Celt probably had yellow or red hair, blue or grey eyes, and a fair complexion.

No satisfactory evidence has been produced that Palaeolithic man occupied Cornwall, but the traces of Neolithic man at the stage when he became acquainted with the use of bronze are abundant. By him were erected the rude stone monuments that are scattered over the county, and he had his favourite sites for trimming flints into scrapers and arrow-heads. One of the most notable of these was the shore of Dozmare Pool.

Of the rude stone monuments the dolmen or cromlech is sepulchral, the dolmen when large having been a tribal or family mausoleum, and the kistvaen, which is far smaller, contained the bones of one individual alone. The dolmen is composed of three or more upright stones sustaining one or more coverers, and was often buried under a cairn. The finest in Cornwall are Zennor and Lanyon Quoits, the Trethevy Stone, and Chûn Quoit.

Stone circles are numerous. Their purpose has not been determined, but on Dartmoor, where they have been examined, they exhibit a floor strewn with charcoal. They have, moreover, usually numerous cairns or barrows associated with them. The finest sacred circle is that of the Stripple Stones on the Bodmin moors, but this has of late years been sadly mutilated. The " Merry Maidens," in St Buryan's parish near Penzance, is a magnificent circle. Other circles are those of Boscawen-ûn, Dauns Môr, Wendron, the Trippet Stones on Blisland moor,

Lanyon Cromlech

Fernacre and Stannon on the St Breward moors, the Hurlers near the Cheesewring, the Nine Maidens in St Columb Major, Duloe, etc.

Menhirs or "longstones" are upright monoliths, probably set up in memorial of the dead. Of these there are many in Cornwall but none of great height. The Pipers in St Buryan are the loftiest, 13 ft. 6 in. and 15 ft.

The Merry Maidens, St Buryan

Of stone alignments Cornwall is almost wholly barren, but one at St Breock can be claimed with confidence. This is the more remarkable as they abound on Dartmoor. But the reason probably is that the stones have been carried off to serve as gateposts, and in some cases are embedded in walls of fields. They were probably erected in commemoration of the dead and are always associated with cairns and interments.

Subterranean chambers, constructed of upright stones with coverers, were possibly store chambers for grain. The best preserved is at Trelowarren.

Upright holed stones are met with in Madron, St Buryan, St Just, Sancreed, Constantine, Wendron, etc. Their purport is unknown.

Very curious are the clusters of communal huts at Chysauster, Bosporthennis, etc. They probably belong to the Iron Age, whereas the hut circles scattered over a hill side, or within a pound, pertained to the early Bronze Age.

Arrow heads, lance heads and scrapers have been found in tolerable abundance on the Bodmin moors, on Carn Brea, at St Agnes, etc., and celts (axe heads) of greenstone and diorite have occurred, but not with great frequency.

At Harlyn Bay has been found a cemetery of the Iron Age, all the bodies in slate cists, crouching. In the cairns and kistvaens (stone coffins) on the other hand the bodies have been burnt. Numerous urns of the well-determined Bronze Age type have been recovered from cairns.

The camps of stone and earth in Cornwall are very numerous. They all—or nearly all—date back to the same period of the early bronze, but may have been used by later peoples. They are of two descriptions, the cliff castles, where a headland is protected by banks and dykes on the side of the mainland, and circular or oval camps crowning heights, with concentric rings of circumvallation. Where the hill top does not admit of the circular form the earthworks adapt themselves to the contour of the hill.

Roman remains are conspicuously rare in Cornwall. Some fragments of Samian ware, coins, and a bronze and silver metal vessel have been found in Bossens, a camp in St Erth, and the head of an ensign at St Just. A second Roman camp is Tregeare, near Bodmin. An inscribed milestone of the time of Constantine the Great is in the churchyard of St Hilary. The metal bowl at Bossens was inscribed by Aelius Modestus to the god Mars.

Of the Celtic period, gold lunettes have been found at Harlyn ; a gold cup near the Cheesewring in a cairn along with a corroded iron weapon ; a portion of a gold armlet at Penzance and of a brooch at the Lizard.

Of Saxon remains the principal are the hoard at Trewhiddle, a silver chalice, finger-ring, pins, etc. Coins have been found; among them one of Ethelred, struck at Launceston. On the altar slab formerly at Treslothan, now supporting a sundial at Pendarves, is inscribed the Saxon name of Ægured; and an old bell at Lanhydrock has on it "Æthelstan sumpta an[ima] sua." These are scanty remains, fewer even than the Roman.

Cornwall is, however, rich in Romano-British inscribed stones, dating from the eighth century down. At St Cleer is the memorial stone to Doniert (Dungarth) son of

Caradoc. In Lewannick churchyard are stones with inscriptions not only in Latin characters but also with Ogams at the angle, showing that the Irish had settled there. Some of the inscribed stones are certainly earlier than the eighth century, to which cautious antiquaries have brought them down.

St Cleer: Monument to Doniert, son of Caradoc, died A.D. 872

Crosses of granite are common, and some are very early, certainly earlier than Athelstan's passage through Cornwall in 938. But others are much later. They vary considerably in size and in shape. Some were used as preaching stations before churches had been built, but others marked the tracks over the moors, and some may have indicated boundaries. Some are excessively

Mawgan Cross

rude, some have the figure of the Saviour carved upon them, and these are comparatively late. Others—as that of St Neot—have elaborate scroll-work on them like those in South Wales.

St Buryan Cross

21. Architecture: (*a*) Ecclesiastical.

A preliminary word on the various styles of English architecture is necessary before we consider the churches and other important buildings of our county.

Pre-Norman or, as it is usually, though with no great certainty termed, Saxon building in England, was the work of early craftsmen with an imperfect knowledge of stone construction, who commonly used rough rubble walls, no buttresses, small semi-circular or triangular arches, and square towers with what is termed "long-and-short work" at the quoins or corners. It survives almost solely in portions of small churches.

The Norman Conquest started a widespread building of massive churches and castles in the continental style called Romanesque, which in England has got the name of "Norman." They had walls of great thickness, semi-circular vaults, round-headed doors and windows, and massive square towers.

From 1150 to 1200 the building became lighter, the arches pointed, and there was perfected the science of vaulting, by which the weight is brought upon piers and buttresses. This method of building, the "Gothic," originated from the endeavour to cover the widest and loftiest areas with the greatest economy of stone. The first English Gothic, called "Early English," from about 1180 to 1250, is characterised by slender piers (commonly of marble), lofty pointed vaults, and long, narrow, lancet-headed windows. After 1250 the windows became broader, divided up, and ornamented by patterns of tracery, while in the vault the ribs were multiplied. The greatest elegance of English Gothic was reached from 1260 to 1290, at which date English sculpture was at its highest, and art in painting, coloured glass making, and general craftsmanship at its zenith.

After 1300 the structure of stone buildings began to be overlaid with ornament, the window tracery and vault ribs were of intricate patterns, the pinnacles and spires loaded with crocket and ornament. This later style is known as "Decorated," and came to an end with the Black Death, which stopped all building for a time.

With the changed conditions of life the type of building changed. With curious uniformity and quickness the style called "Perpendicular"—which is unknown abroad—developed after 1360 in all parts of England and lasted with scarcely any change up to 1520. As its name implies, it is characterised by the perpendicular arrangement of the tracery and panels on walls and in windows, and it is also distinguished by the flattened arches and the square hoods over the doorways, by the elaborate vault-traceries (especially fan-vaulting), and by the use of flat roofs and towers without spires.

The mediaeval styles in England ended with the dissolution of the monasteries (1530—1540), for the Reformation checked the building of churches. There succeeded the building of manor-houses, in which the style called "Tudor" arose—distinguished by flat-headed windows, level ceilings, and panelled rooms. The ornaments of classic style were introduced under the influences of Renaissance sculpture and distinguish the "Jacobean" style, so called after James I. About this time the professional architect arose. Hitherto, building had been entirely in the hands of the builder and the craftsman.

Cornwall does not furnish stately and richly adorned churches as does Devonshire, and even more so, Somersetshire. This is due to the intractable material available, granite, which unlike that of Brittany did not lend itself to rich sculpture. Only the Elvan stone, already described, could be worked with delicacy, and this is easily corroded by the weather, and the Catacluse stone is black. The beautiful Polyphant stone of Lewannick does not seem to have been largely employed. Although so close to Launceston, when the church there was reconstructed by Sir Henry Trecarrel, in the reign of Henry VIII, he employed only granite, which was sculptured with infinite labour, and with poor effect.

Buried Church, Perranporth

The most ancient churches in Cornwall were probably exceedingly rude. Only two of the earliest remain and these are in ruins, Perranporth and St Gothien, very much resembling churches of the most primitive period in Ireland.

There are some important remains of Norman architecture, notably the west front and part of the nave of St Germans, some portion of Blisland, the south door of Kilkhampton, the westernmost arches of Morwenstow, Tintagel, and some tympana of doorways with

Norman Doorway, St Germans Church

rude sculpture. In many of the churches the early fonts remain. Originally baptisms took place in tanks or barrels sunk in the earth; then, when adult baptisms ceased, square or circular troughs were placed on the floor, and later were raised on pedestals; and not infrequently the pedestal is an addition, at a later date, to the basin of the font itself.

Of the Early English period, the remains are scanty. St Anthony in Roseland is the most perfect example in Cornwall.

Tympanum, Egloskerry Church

Of the Middle Pointed, or Decorated period, are Padstow, St Columb Major, Sheviock, Lostwithiel—where the very peculiar spire is singularly beautiful and foreign in character—and portions in Lanteglos by Fowey, and St Ive near Liskeard.

But, as in Devonshire, so in Cornwall, there was an outburst of church building in the Third Pointed or Per-

pendicular period. Then only—since the early sculptors
of the crosses—did architects and carvers attack the granite,
and most of the churches were then rebuilt. The finest
are St Kewe, Mawgan in Pyder with a beautiful tower,
Probus with a richly sculptured tower of the reign of
Elizabeth, St Austell—also with a good tower, the carvings
in Elvan—St Neot, Stratton, and Buryan. One remarkable

Lanteglos Church

feature is in the porches, where tracery of an ogival
character is introduced in the arch. The Rev. W. Haslam,
in the *Transactions of the Exeter Dioc. Arch. Society*, says
of the Cornish churches that they "are low, and somewhat flat in the pitch of the roof, and without buttresses
to break the long plain horizontal lines which are so
conspicuous. All these are features of the Perpendicular

style, I admit, but not to the extent to which they are carried in Cornwall.   Besides this, the general form of a Cornish church is plain; externally, the plan of the larger ones is a parallelogram, divided into three low ridges of roof: there is a porch on the south side; this is the only break in the horizontal line I allude to.   The smaller churches have generally but one aisle, and these have a

Launceston Church

transept also, but sometimes two transepts; but even these do not relieve the plainness of the exterior.   This is not the character of one church, or two, or three; but more or less of all.   It is their character, and I attribute it to the boisterous nature of the climate in that narrow county, exposed as it is, with very little shelter, to violent storms from the sea on both sides."

No great elaboration of tracery was possible with granite, and the architects abandoned the thought of making the churches attractive externally, devoting their attention to the internal decoration. The appearance of a large Cornish church now is that it is a long low shed, lacking in height and dignity. But the architect divided it in two by the screen, and thus brought it into proportion. Unfortunately, however, in very few have the beautiful rood-screens been left, which were generally spared in the Devonshire churches. Those of Cornwall in no way fell short of those in Devon, but the Puritans first of all, and then the barbarians of the Georgian period swept them away, and the churches in the nineteenth century fell into the hands of local architects who left them "naked, swept"—but not "garnished." They were, let us hope, the last of the Cornish wreckers. A few, but only a few screens remain.

In the interior of the Cornish churches the chief feature is the absence of a chancel arch, which is almost universal. The arch was unnecessary when the roodloft extended upwards and was backed by a painted board. In some of the churches there are interesting bench ends. At St Austell, the miners' tools are represented on them; on some rabbits running in and out of their burrows are figured, and seagulls are frequent.

In painted glass Cornwall is not rich, except in the 15 windows of St Neot near Liskeard, and in that in the north aisle, and the fragments in the south aisle of St Kewe.

The church towers in Cornwall are for the most part square without buttresses, and with four pinnacles. One peculiarity of these pinnacles is that they are often cut on a curve to lean outwards, probably to mitigate the stiffness of effect.

A few words must be devoted to the new Cathedral

Landewednack Church

at Truro erected from the designs of the late Mr J. L. Pearson in a French Gothic style of architecture.   It con sists of a choir of five bays, with retrochoir, transepts, nav e with north and south aisles, baptistery and south- we st porch, richly sculptured ; a central and two western towers. Attached to the south side of the choir is the fine old parish church of St Mary, or a portion of it. There is an elaborately sculptured reredos over the altar. Notwithstanding its faults Truro Cathedral is a notable

Dupath Well, Callington

monument of the enthusiasm and self-denial of the Cornish people.

The Holy Wells of Cornwall are a distinct feature of the county. In all Celtic lands, previous to the introduction of Christianity, there was a great veneration for wells, and the early missionaries took advantage of this to turn them into baptisteries, or in other ways to consecrate them. Holy wells abound in Cornwall, but they have not always much architectural character. That of Dupath, by Liskeard, is the finest, but there is also another that is fine at St Cleer, and one most interesting and unique at St Clether, where, indeed, there are two, for the water from the first flows into a chapel and is carried under the old stone altar, to decant into another well outside the chapel. The Madron Holy Well was for long famous for cures.

22. Architecture: (*b*) Military—Castles.

Before the Norman Conquest there were no masonry built castles in Cornwall, only stockades of wood surmounting earthworks or piled up masses of stone uncemented. The usual Saxon *Burh* was a mound, surmounted by a structure of timber, reached by a bridge or ladder from a base-court that was encompassed by moat and mound and stockade. The Norman system of building a castle was to erect a round or square keep, a massive structure of stone, on the mound that had formerly been surmounted by a wooden structure, and to surround the base-court with a stone wall. Within this were erected the necessary domestic buildings. Very generally the entrance to the court was strongly defended by a second tower. The style of castle was greatly altered in the reign of Edward I but of such Edwardian castles there are no examples in Cornwall, save the poor fragment of Tintagel. The Normans built a castle àt Launceston, and there the circular keep standing on a lofty tump of rock, artificially shaped, is of their construction, but the ruined buildings below, with the gateways and walls of the base-court, are later.

The castle of Trematon also consists of a "motte" surmounted by a circular keep, and a base-court with square tower at the entrance, with an archway.

Tintagel Castle is reduced to a miserable ruin, part on the mainland, part on the islet, with the intervening portion blocked up by fallen rocks, forming a narrow isthmus. The deep chasm that formerly separated the two portions of the castle was anciently spanned by a drawbridge. The work appears to be of the thirteenth century. On the island are the remains of a chapel with its altar slab still in place. Tintagel became a residence of the Earls of Cornwall, and in 1245, Richard son of King John received in it his nephew David, Prince of Wales. It was subsequently used occasionally as a prison. In the reign of Elizabeth, that penurious queen, deeming the expense of keeping it up too onerous, allowed it to fall into ruin.

St Michael's Mount was crowned with a castle and a church. The oldest portion is the central tower, of the fourteenth or fifteenth century; other portions are later additions, and much very bad modern work has tended to its sad disfigurement. Edward the Confessor planted a monastery on the rock, and granted it to Mont S. Michel

in Normandy; at the Conquest it was made over to Robert
Earl of Mortain, but the Benedictines of Mont S. Michel
continued to have a cell there with a prior.   It was consequently at once a religious house and a military post.

Restormel, near Lostwithiel, consists of a keep crowning a hill, with a gatehouse on the west, a projecting
tower on the E.N.E., and a chapel.   It is not older than

St Mawes Castle

the reign of Henry III and was the stronghold of the
Cardinhams and then of the Traceys, from whom it
passed to the Earls of Cornwall.   The circular keep is
only 30 ft. high.   The castle was already ruinous in the
time of the Civil War, but it was put in repair and held
by the Parliamentary forces till taken by Grenville.

St Mawes is a small but perfect castle, erected by
Henry VIII.

Pendennis Castle was another erection of Henry VIII, on the site of earlier fortifications. The circular tower dates from his time, but it was added to considerably in the reign of Elizabeth. In 1644 Pendennis afforded shelter to Queen Henrietta Maria, when embarking for France, and hither came Prince Charles in 1646 on his way to Scilly.

Helsborough, near Michaelstow, was a fortress belonging to the Earls of Cornwall, but it shows no tokens of having ever been walled with masonry. The only structural remains to be seen are the ruins in the midst of a Perpendicular chapel.

On Carn Brea is a tower, another on Roche rock; and St Catherine's Castle, erected by order of Henry VIII, defended the entrance to the harbour of Fowey.

23. Architecture: (c) Domestic and Monastic.

In England generally castles belonged only to the Crown or to great nobles, and no gentleman was suffered to castellate or embattle his walls without a special licence from the Crown. In Cornwall all the castles pertained to the Crown or the Duke of Cornwall, and private persons had to content themselves with purely domestic mansions. Till the reign of Elizabeth the dining-hall reaching to the roof was the most conspicuous feature, and opening out of it was the ladies' bower, a small oak-panelled room. The inconvenience arising from a house being cut in half by the hall led in the reign of Elizabeth to an alteration, and the halls were ceiled over, so that the upper portion could be used for bedrooms and passage. Before her reign the usual form of a house was quadrangular, that is to say a court surrounded by buildings entered by a gate, with the hall and principal portions of the house opposite the entrance gate. But in the reign of Elizabeth it became the fashion to form the house in the shape of the letter E. In her father's reign it often had the shape of the letter H with the open ends closed by slight walls.

Cornwall possesses very few stately houses. At the close of the seventeenth century a schoolmaster at Trebartha filled a folio with sketches of the ancient manor-houses of the neighbourhood of the Tamar; picturesque old mansions of the reigns of Henry VIII and Elizabeth. Nearly every one has disappeared. The squirearchy of Cornwall, flush of money, through tin, pulled down their old residences and built mansions in the Georgian period, totally devoid of interest. Of the old houses few remain except as farm houses. They were, however, never so magnificent as those in the counties where bricks and easily dressed stone existed. But still there remain Cothele, unaltered, the beautiful house of Lord Mount Edgcumbe on the Tamar; Basil, the manor-house of the Trevelyans, much mutilated by the barbarous hand of a modern architect; Trecarrel, near Launceston, an old Tudor mansion with a noble hall, never completed; Place, near Padstow, formerly Prideaux Castle, the Elizabethan residence of the Prideaux Brunes, very stately, and with a dining room rich in carved oak; Lanherne, built in 1580, a small manor-house of the

Arundells and now a convent; Lanhydrock, built by the
first Lord Roberts—they called themselves later Robartes;
Place House, Fowey, with rich sculpture; Trerice, the
old seat of the Arundells; Tregudick with its Elizabethan
hall; Penheale, once the seat of the Speccots, in Egloskerry;
Tonacombe, an unaltered house of the reign of Henry VII,

Cothele

in Morwenstow; Penfound in Poundstock, small but
charming; and Lanreath, with a carved oak parlour, the
ancient house of the Grylls family. There are others,
now farmhouses, and only spared on that account, deserted
when the squires moved elsewhere and did not pull down
their ancestral residences.

A monastic building consisted of a church, with a

cloister court adjoining, about which were the dormitories, a library, and a refectory or room in which all had their meals in common. There would be often two courts, one outer, the other with the cloister about it in which the monks mainly lived and in the centre of which was the monastic graveyard. The garden of pot herbs and herbs for medicinal purposes was an essential feature of all monastic settlements, as was also the stewpond or ponds for fish. The necessity under which the monks lay of being near water, both for their fish and for sanitary purposes and for drinking, led to all monastic establishments lying low down in valleys by running streams.

Of monastic remains there are few. At Launceston the foundations of St John's Priory have been laid bare. At Glasney in Penryn a few walls alone represent what was once a stately priory. Of the great house at Bodmin hardly a wall stands, but some remains of the sanatorium exist at Lavethan. At Lanivet are the remains of St Benet's Monastery, till 1859 the most picturesque and best preserved of the monastic buildings in Cornwall, except St Michael's Mount. An engraving of the remains was published by Lysons in 1814, which shows the house to have been beautifully situated, and as beautiful as its situation demanded. At the date mentioned it was mutilated and spoilt.

Cornwall cannot boast picturesque cottages. Some few remain that possess some charm, as the old Post-office at Tintagel, some slate-hung dwellings in West Looe, the Lugger Inn, Fowey, and the almshouses at St Germans. There are as well some that call for an artist to use his pencil at Saltash. But, on the whole, the county is poor in the domestic architecture of farm and cottage, and house fronts in the towns, with rare exceptions, are not of any artistic character.

The Old Post Office, Tintagel

24. Communications—Roads, Railways.

The great road from London to Falmouth ran through Launceston, Bodmin, and Truro, and was kept in good order, and over it raced the mail coaches that conveyed letters on to the packets at Falmouth. This was the main artery of communication for more than a century, till Falmouth was abandoned as a mail-packet station by the Government. From 18 to 20 fine vessels performed this service, carrying letters and papers to all parts of the world, until the extension of railways caused the service to be transferred to Southampton.

Whether there were systematically constructed Roman roads in Cornwall has been doubted. One curious ancient road—the Giant's Hedge—is found near Lanreath, and appears to have been a portion of a road raised on a bank that started from a ferry over the Tamar and was carried into the west of Cornwall. There was a road also that came from Exeter and crossed the Tamar at Polson Bridge and then turned north to Camelford. Another ran past Stratton to the estuary of the Camel opposite Padstow, where Romano-British remains have been found on Bray Hill. But it is possible enough that these roads were of British and not of Roman construction.

In the Middle Ages little or nothing was done to keep the roads in repair. Even in the eighteenth century all that was thought necessary was to throw down a load of boulders into the ruts, rake them in, and leave coach and cart wheels to grind them up. But the roads were taken in hand at the beginning of the nineteenth century, and the turnpikes served to provide funds for keeping them in order, while Macadam's invention of breaking the stones before laying them on the roads assisted enormously in facilitating transit along them.

The roads in Cornwall are now for the most part excellent, metalled with elvan, and cyclists and motorists can have little to complain of on that score.

Two railway companies have penetrated the county; the G.W.R. in the south crosses the Tamar by the famous Saltash tubular bridge, and runs to Penzance,

A Cornish Stile

by Truro. There is a branch to Bodmin, another to Fowey, and another crosses the county to Newquay. After reaching Truro, the line deserts the south, but sends a branch down to Falmouth. It runs to Redruth and Camborne and reaches the sea at Hayle. It sends a short branch up to St Ives, but the main line turns south again to end in a terminus at Penzance. The L. & S.W.R., after sending out a branch to Bude, reaches Launceston, and then supplies the dreary country from Launceston to Camelford with communication. From Camelford it runs to Wadebridge; and down the estuary to Padstow, a branch to the south serving Bodmin.

In addition to the branches already mentioned are others to Looe and Helston.

Penzance is reached from Paddington in $8\frac{1}{2}$ hours and Newquay in half-an-hour less.

The L. & S.W.R. leaving Waterloo reaches Launceston in $5\frac{1}{2}$ hours, Bude in 6 hours, and Wadebridge in under 7 hours. From Wadebridge coaches run to Newquay.

A steamer maintains communication with the Scilly Isles from Penzance.

25. Administration and Divisions, Ancient and Modern.

Before the Conquest the divisions of the county were probably those afterwards forming the old deaneries, and followed the limits of the Celtic tribes under their several chiefs. Of these there were eight :—East and West, Kerrier, Penwith, Powder, Pyder, Trigg Major, and Trigg Minor, but at the Conquest a redistribution was made in hundreds. These were Conarton, Fawiton, Pawton, Riatton, Stratton, Tibesta or Tibesterna, and Winneton. There may have been an Anglo-Saxon redistribution. But it was a rearrangement that did not last, and never commended itself to the people, and it is not easy now to ascertain what the limits were. Conarton was Penwith. Perhaps Tibesta was Powder and Winneton Kerrier. East and West composed one district of Wyvelshire.

Cornwall was first an Earldom, and then accordingly a County, but when raised to be a Dukedom it became a Duchy. It had two chief officers, the Earl and the Sheriff, the latter appointed by the crown.

The county was divided up into hundreds for administrative purposes. Each hundred was supposed roughly to contain a hundred free families. Each hundred had its own court, and every township its assembly under the reeve. But the tinners were under their own laws and officers, and their court, called the Stannary Court, sat formerly at Lostwithiel. Every manor also had its court.

All the hundreds of Cornwall, except Penwith, from time immemorial belonged to the Earls of Cornwall. The hundreds and Petty Sessional Divisions are now coextensive, and are as follows:—1. Hundred of East (Northern Division). 2. Hundred of East (Southern Division). 3. Hundred of Kerrier. 4. Hundred of Lesnewth. 5. Hundred of Penwith. 6. Hundred of Powder, East.

7. Hundred of Pyder, West.   8. Hundred of Pyder,
East.   9. Hundred of Stratton.   10. Hundred of Trigg.
11. Hundred of West.

Cornwall formerly returned four county members in
two divisions, but under the provisions of the "Redistribution of Seats Act, 1885," it now returns six members
in six divisions.   1. The western or St Ives division.
2. The north-western or Camborne division.   3. The
Truro division.   4. The Mid or St Austell division.
5. The south-eastern or Bodmin division.   6. The north-
eastern or Launceston division.   Under the provisions
of the above-mentioned Act, the boroughs of Bodmin,
Helston, Launceston, Liskeard, St Ives, and Truro were
deprived of independent representation and merged in the
county, and Penryn and Falmouth lost one member.

In 1877 the diocese of Truro, taken from that of
Exeter, was formed, comprising the whole county of
Cornwall, together with the parishes of Broadwood-
Widger, Virginstow, Werrington, St Giles on the Heath,
and North Petherwin, which are in the county of Devon.
The Stannary Court is now at Truro, but its occupation
is almost gone.

The Poor Law Unions are Bodmin, Camelford,
Falmouth, Helston, Launceston, Liskeard, Penzance,
St Austell, Redruth, St Columb, St Germans, Scilly Isles,
Stratton, and Truro.

The County Council formed under the Local Government Act of 1888 consists of a chairman, aldermen, and
councillors; but for the local government of the towns
and parishes another Act was passed in 1894, and new

The Old Guildhall and Pillory, Looe

names were given to the local bodies.   In the large urban parishes the chief authorities are now entitled District Councils, while the smaller parishes have their Parish Councils or only Parish Meetings.

The county is in the western circuit; the assize and quarter sessions, which were formerly held at Launceston, a most inconvenient place for the purpose, being at the extreme limit of the county, are now held at Bodmin.

In Cornwall there are 223 civil parishes and the municipal boroughs are eleven, Bodmin, Falmouth, Helston, Launceston, Liskeard, Lostwithiel, Penryn, Penzance, Saltash, St Ives, and Truro.

The civil parishes and those that are ecclesiastical are not always conterminous.   Of the latter there are 236. There are two Archdeaconries, Cornwall and Bodmin, and twelve deaneries, St Austell, Carnmarth, Kerrier, Penwith, Powder, and Pydar in the Archdeaconry of Cornwall, and Bodmin, East, Stratton, Trigg Major and Minor, and West in that of Bodmin.   There is a Bishop at Truro and a suffragan who takes his title from St Germans.

26. Roll of Honour.

It would, perhaps, be invidious to say that Cornwall has produced men of more brilliant and varied achievements than any other county in England, but she can certainly show a very notable roll of honour. As might be expected from her geographical position, aided by good harbours, she has produced some great seamen who have done gallant service for England. At the head of these must come Sir Richard Grenville, hero of the *Revenge*, whose action off the Azores in 1591 has rendered him one of England's immortals. Trapped by the huge Spanish fleet off Flores, Sir Richard had many of his crew sick on shore, but declined to leave till they had been brought on board. The *Revenge* engaged fifteen large Spanish men-of-war and stood at bay from three in the afternoon all through the night till the following morning, when the last barrel of powder was spent. Ralegh told of it, as did Gervase Markham in 1595, and Tennyson nearly 300 years later in the stirring lines :—

" Ship after ship, the whole night long, their high-built galleons came,
Ship after ship, the whole night long, with her battle-thunder and flame ;
Ship after ship, the whole night long, drew back with her dead and her shame.
For some were sunk and many were shattered, and so could fight us no more—
God of Battles, was ever a battle like this in the world before ? "

His grandson, Sir Bevil Grenville, "the Mirror of Chivalry," was a scarcely less notable warrior on land and fell in the Royalist victory of Lansdowne near Bath in 1643. Admiral Edward Boscawen, "Old Dread-

Captain Bligh

nought," third son of Viscount Falmouth, distinguished himself at the taking of Cartagena and in the Cape Breton expedition, but his most memorable deed was the defeat of the French Toulon fleet in Lagos Bay in 1759. Captain Bligh, noteworthy as the captain of the *Bounty*, of which the mutiny is one of the most familiar tragedies of the sea, was perhaps in great measure the author of his own misfortunes, for he was a man of very overbearing temper, but his journey of 3618 miles in an open boat after having been set adrift with others by the mutineers was a remarkable feat. The Pellew family has added at least two names to the roll of distinguished sailors. Edward Pellew, when in command of the *Nymphe*, manned by Cornish miners, captured the French man-of-war *Cléopâtre* under peculiarly gallant circumstances, the first of a series of brilliant exploits which led to his being created Baron Exmouth. Later, in 1816, he bombarded Algiers, reduced the Dey to submission, and put an end to the Barbary corsairs. His brother, Admiral Sir Isaac Pellew, commanded the *Conqueror* at the battle of Trafalgar.

Cornwall being a land of mines has developed machinery and furthered invention in this direction. The most notable of the inventors she has produced is Richard Trevithick, who first made the high-pressure engine, and is still more remarkable as the early pioneer of motor traffic, putting his road locomotive on the Camborne highway on Christmas Day, 1801, and obtaining a speed with it of 12 miles per hour. In 1812 he laid before the Navy Board his invention for a screw propeller for ships, only to meet with a refusal. Sir Goldsworthy Gurney, born in 1793, also ran steam-motors on the roads until they were forbidden by Act of Parliament, and the work of a lifetime and his fortune of £30,000 vanished into thin air. The oxy-hydrogen blowpipe and the steam-jet were invented by him. The wreck of H.M.S. *Anson* on Looe bar with the loss of over 100 lives in 1807 had a great effect on one of the spectators, Henry Trengrouse, a Helston cabinet-maker, who thereupon invented the

Sir Humphry Davy

rocket life-saving apparatus and spent £3000—all his means—in experiments and in vain endeavours to induce Government to adopt the system. Another great benefactor to mankind was Sir Humphry Davy, the son of a poor gilder near Penzance. His safety-lamp he nobly refused to patent lest the sphere of its usefulness should be restricted, and he is fittingly honoured with a statue in his native town.

More than one distinguished traveller finds place among Cornish worthies. Richard Lander, the son of a Truro innkeeper (1805–34), stands at the head of them. He went with Clapperton to Sokoto and on his death took up his work, tracing the mouth of the Niger on a second expedition, and dying on a third at Fernando Po. Peter Mundy, born about 1596 at " Penrin, a pretty towne in Corne Wall," as he describes it, was one of the most remarkable travellers that the West of England has produced, whether in virtue of his long trading voyages to the Far East, or of his continental wanderings, of which he kept a not less careful record. James Silk Buckingham, who died in 1855, wrote eighteen books of travel, but was mainly noteworthy for his endeavours to do away with the monopoly of the East India Company.

Among statesmen must be noticed Sir John Eliot, born at Port Eliot in 1592, and at one time friend of George Villiers, Duke of Buckingham, with whom he later broke, and taking up a strong line in the Parliament of 1628 against arbitrary taxation helped to force the Petition of Right from Charles. He was ultimately committed to the Tower, where he died three years later in 1632. To give an adequate sketch of the life of Sidney Godolphin would be to give the history of the times of Charles II, James II, William of Orange and Anne, with each and all of whom he was closely implicated. An extraordinarily able financier, his character was such as to permit him to serve either party indifferently. "Sidney Godolphin," said Charles, "is never in the way and never out of the way." Lord High Treasurer under Queen Anne, he was made Earl of Godolphin, only to be disgraced in 1710 and die shortly after.

Samuel Drew, the "Cornish Metaphysician," who was born at St Austell, had been a smuggler and a shoemaker in his earlier days, but developed into a Wesleyan preacher and became the author of an essay on the Immortality of the Soul. Of more normal mould was Humphrey Prideaux, born at Padstow in 1648, who wrote a *Life of Mahomet*, and *The Old and New Testament Connected*, which reached its 27th edition only a few years ago. He became Dean of Norwich and died in 1724.

Cornwall has produced several local antiquaries, as the Rev. Richard Polwhele, who died in 1838; the Rev. William Borlase, d. 1772; William Hals, the historian of the Duchy; and William Sandys, who died in 1874. She has had but few true poets, but the Rev. Robert Stephen Hawker, though not actually born in Cornwall, spent all his life there and may certainly come under this heading. He became Vicar of Morwenstow in 1834, and remained there till his death in 1875, having during 'this time transformed his parishioners from a set of lawless wreckers to a decent community. His poems were almost all connected with Cornish subjects, and one of the best known of them is that on Bishop Trelawney's imprisonment—"A good sword and a trusty sword," with the refrain—"And shall Trelawney die?

There's twenty thousand Cornishmen ; Shall know the reason why."

   " A wondrous Cornishman, who is carrying all before him " is Sir Joshua Reynolds' description of John Opie the painter, who was born the son of a poor carpenter

John Opie

near Truro in 1761, became an R.A. and portrait-painter of great note, and died in 1807. His second wife was Amelia Opie the novelist, daughter of the Norwich physician, Dr Alderson. Henry Bone, the enamellist, is also of sufficient distinction to deserve mention here.

Of the astronomers of modern times few have attained the eminence of John Couch Adams, the discoverer of the planet Neptune, who was born at Laneast in 1819. As a shepherd boy he loved to lie on his back and watch the stars, and he at once devoted himself to the study of astronomy when he was sent to school at Saltash. He became Senior Wrangler at Cambridge in 1843, and soon after taking his degree, being struck with irregularities in the motion of Uranus, he made a series of calculations and observations which resulted in the discovery of the new planet, the French astronomer Leverrier having simultaneously recorded its existence. A distinguished geologist and one of the pioneers of scientific cave exploration was William Pengelly, whose great work was the thorough examination of Kent's Cavern near Torquay, a labour which lasted from 1865 to 1880. Last, but by no means least worthy of mention in our list, must come Davies Gilbert, a great discerner of rising genius, to which he was ever ready to lend help and encouragement. The patron of Davy, Trevithick, Hornblower, and Goldsworthy Gurney, he was himself a man of remarkable and varied abilities, scientist, mathematician, and antiquary, and President of the Royal Society. He died in 1839.

27. THE CHIEF TOWNS AND VILLAGES OF CORNWALL.

(The figures in brackets after each name give the population in 1901 and those at the end of each section are references to the pages in the text.)

Bodmin (5353), now held to be the county town. Formerly there was a priory here, founded in 938. The church, the largest in Cornwall, is Perpendicular, except the tower and a part of the chancel, which are earlier. In it is the fine monument of Prior Vyvyan (1533). Bodmin is a municipal borough, market, and union town, and head of a county court district. The Assizes are held here. The town is pleasantly situated nearly in the centre of the county. A branch of the G.W.R. leads from Bodmin Road station on the main line. The L. & S.W.R. has also a branch to Bodmin from Wadebridge. The prison stands about half a mile north-west of the town; and the County Lunatic Asylum a little to the west. Bodmin has also a Hospital and Dispensary, and Barracks for military. (pp. 9, 14, 18, 24, 67, 96, 102, 104, 111, 131, 132, 135, 137, 139.)

Bude (see Stratton).

Callington (1714), a small market town, formerly a parliamentary borough returning two members. Callington is in the ecclesiastical parish of North Hill, and the church is merely a chapel of ease. It is in the Perpendicular style, and has the unusual feature in Devon and Cornwall of having a clerestory. The L. & S.W.R. has constructed a light railway from Beer Alston to Callington and Calstock. (pp. 105, 124.)

Calstock (5874) is a large village and township. The church is at a distance from the village or town. Gunnislake, two miles north of Calstock, is in the parish. Both places have declined in numbers and prosperity through the abandonment of the mines in the neighbourhood. (pp. 15, 69, 72, 87, 101.)

Camborne (14,726), a market town with a station on the main line of the G.W.R. It is a great seat of mining operations. (pp. 38, 81, 101, 135, 137.)

Camelford (1384), a small market and union town, in a bleak and wind-swept situation, consisting of one street. It is the head of a county court district and magisterial division. It is in the parish of Lanteglos, nearly two miles distant, and has not in it even a chapel of ease. Camelford was an ancient borough with mayor and corporation. (pp. 14, 105, 133, 135, 137.)

Falmouth (3207), a seaport, market, and union town and a municipal borough. It is the head of a county court district. Falmouth was incorporated by charter 1661. It has the advantage of possessing one of the finest and most capacious harbours in the country. As many as 350 to 400 sail have taken refuge here in winter at various times. There are docks, factories, and building yards. Of late an attempt has been made to turn Falmouth into a winter resort for invalids, and the coast is advertised as "The Cornish Riviera," but it is a Riviera without the sun of the Mediterranean. (pp. 30, 31, 48, 66, 88, 92, 94, 95, 96, 132, 135, 137, 139.)

Fowey (2258), once a borough, is a market town and a shipping port with a station on a branch of the G.W.R. The church of St Finbar is a noble edifice in the Perpendicular style but with portions on the north dating from 1336. Place House, the seat of the Treffry family, is a beautiful, richly-sculptured

Fowey

mansion of the reign of Henry VII.   Fowey consists of one very
narrow street.   The town was created a borough by Elizabeth,
returning two members, but was disfranchised in 1832.   (pp. 49,
88, 105, 119, 128, 130, 131, 135.)

Grampound (491) deserves mention only as having returned two members to Parliament before the Reform Bill, and
as formerly a borough.   (pp. 18, 105.)

Helston from Bullock Lane

Helston (3088) is a municipal borough, market, and union
town, and the resort of all who want to buy or sell in the Lizard
district.   It was not formed into an ecclesiastical parish till 1845.
A branch of the G.W.R. reaches it.   Below the town is the
beautiful Loe Pool.   The principal streets form a cross, and have
a constant flow of water through them.   (pp. 8, 104, 135, 137, 139,
143.)

Launceston (14,310), the ancient capital, is a municipal
borough, market, and union town, and head of a county court

district. The town occupies a height, and above it towers the ancient castle, but a portion of the town, Newport, lies in the valley at the foot. Both the G.W.R. and the L. & S.W.R. have stations at Launceston. The church of granite is richly sculptured throughout; but in a debased Perpendicular style. It replaced an older church of which the tower alone remains. (pp. 7, 15, 25, 67, 96, 104, 111, 117, 121, 126, 129, 131, 132, 135, 137, 139.)

Liskeard (4945), a municipal borough, market, and union town, head also of a county court district, with a station on the

Launceston

G.W.R. Liskeard returned two members to Parliament till disfranchised by the Redistribution Act of 1885. The church of St Martin is the largest in the county next to that of Bodmin; it is in the Perpendicular style but retains portions of earlier work. Liskeard mainly flourishes on the granite quarries of the Cheesewring; it did at one time flourish still more on the mines of tin and copper in Caradon. (pp. 12, 18, 31, 104, 119, 122, 125, 137, 139.)

Looe (2588) is composed of two towns, East Looe and West Looe, separated by a creek and united by a long bridge. East Looe was formerly a borough returning two members. It is the terminus of a branch line of the G.W.R. There is a good tidal harbour, to which vessels of large tonnage can be brought up for the shipment of ore and granite, conveyed thither from Caradon and the Cheesewring, and thence coal is conveyed to Liskeard

Looe

and its neighbourhood. The old town hall is a picturesque building. West Looe was also originally a borough returning two members. (pp. 31, 49, 50, 92, 105, 131, 135, 138, 143.)

Marazion (1252) or Market Jew derives its name from a Thursday market there held. In ancient times the place was supported by the pilgrims resorting to St Michael's Mount. The town was pillaged by the French in the reign of Henry VIII. From Marazion St Michael's Mount is reached. At the base of the Mount lies a fishing village of 38 houses, and a little harbour. From the sea the hill rises to a height of 230 ft., the body of the island is granite but its north base is of slate. The castle is the property of Lord St Levan. Marazion is a station on the G.W.R. (pp. 9, 44, 96.)

Mevagissey (2088), a seaport town situated on a fine bay six miles south of St Austell, the nearest railway station. The natural harbour is safe and sheltered. A new harbour has been constructed. The inhabitants depend almost wholly for subsistence on the fishery of pilchards. (pp. 48, 89, 92, 94.)

Newlyn (1302) is not a town, it is an ecclesiastical parish formed in 1848 out of the parishes of Paul and Madron. It has a harbour, and is a seat of fishery. Many artists have been attracted hither by the mildness of the climate and the picturesqueness of the quaint old streets, and of the fisherfolk. (pp. 44, 92, 94.)

Newquay (2935) is a new and rising sea-bathing and holiday resort, to which run two branches of the G.W.R. It possesses a beautiful stretch of sands, and is near some of the finest coast scenery in Cornwall. (pp. 37, 91, 135.)

Padstow (1566) is a quaint old seaport on the estuary of the Camel. It has a fine church with a beautiful carved font. Above the town is Prideaux Place, an Elizabethan mansion. The town lies in a valley and consists mainly of one street. It is a market town, and is connected with the L. & S.W.R. by a branch line. (pp. 35, 55, 56, 57, 67, 92, 94, 119, 129, 133, 135, 145.)

Penryn (3256), a market town and municipal borough, situated on an arm of Falmouth harbour that is being gradually silted up. The parish church is dedicated to St Gluvias. It formerly returned two members to Parliament. Here was formerly a monastic establishment, at Glasney, but of this only a few fragments of wall remain. There is a station on the branch to Falmouth of the G.W.R. (pp. 96, 105, 131, 137, 139, 144.)

Penzance (12,155), the most westerly town in England. The name signifies the Holy Head; it is a seaport, municipal borough, and union town. The climate is here most warm and relaxing. Penzance was incorporated in 1614 and has a corporation consisting of a mayor, six aldermen, and eighteen councillors. There are some fine public buildings, and a good museum for the birds, minerals, and antiquities of Western Cornwall. Hence starts the steamer that communicates with the Scilly Isles. From Penzance excursions are made to the Land's End district, which abounds in magnificent coast scenery, and in prehistoric megalithic monuments. (pp. 44, 57, 66, 88, 96, 108, 111, 133, 135, 137, 139, 143.)

Probus (1207) is a large village rather than a town, and is situated two and a half miles from Grampound station on the G.W.R. It is chiefly remarkable for its church tower, sculptured throughout, and erected in the reign of Elizabeth. (p. 120.)

St Agnes (4291), a market town, reached by a branch of the G.W.R. The church is Decorated, and has a spire, an unusual feature in Cornwall. (pp. 55, 110.)

St Austell (3340), a market and union town with a station on the G.W.R. A place that flourishes on the china-clay works in the neighbourhood. The church has a noble tower and is in the Perpendicular style. It possesses an exceedingly curious and early font. Near St Austell is Menacudle, perhaps the most picturesquely situated and best preserved of the Holy Wells in Cornwall. (pp. 9, 24, 26, 48, 85, 120, 122, 137, 139, 145.)

St Blazey (2931), a market town with a station on the Newquay branch of the G.W.R.

St Columb Major (2640), an ancient market town two and a half miles from the St Columb Road Station on the Newquay branch of the G.W.R. and also in connexion with the L. & S.W.R. at Wadebridge by coach. The church is fine, Perpendicular in style, and has an interesting cross in the churchyard. The chancel, formerly ten feet longer, was wrecked by an explosion of gunpowder stored in the church, in 1676. There is a picturesque old building with state hangings, originally a residence of the Arundells of Trerice, near the churchyard gate. The old Cornish hurling is still practised at St Columb. (pp. 9, 18, 99, 108, 119, 137.)

Cross in Churchyard, St Columb

St Germans (2126), a large village rather than a town, with a station on the G.W.R. It returned representatives to Parliament from 1562 till the passing of the Reform Act in 1832. The church is especially interesting from its noble Norman west front. It lies in the park of the Earl of St Germans. The almshouses in the town are especially to be noted for their picturesqueness. (pp. 105, 118, 131, 137, 139.)

St Ives (6699), a municipal borough on the shore of St Ives Bay, and the terminus of a branch of the G.W.R. from St Erth. The corporation now consists of a mayor, four aldermen, and twelve councillors. The borough formerly returned two members to Parliament. The town is irregularly built with narrow streets and has "a very ancient fishlike smell." The church is interesting, it possesses curious carved bench-ends, and an old cross outside, of an unusually elaborate description. The population is almost entirely dependent on the fishing industry; but of late a St Ives school of painting has grown up, and artists have settled here to study sea effects, and seafaring people. (pp. 38, 39, 67, 87, 88, 92, 93, 94, 102, 105, 135, 137, 139.)

St Just in Penwith (5646), a market town seven miles west of Penzance, the nearest railway station. The town consists of several streets diverging from a triangular space in the centre. The church, well cared for, is Perpendicular in style. (pp. 110, 111.)

St Mary's, Scilly is a town and the focus of life in the Isles of Scilly. The total population of these isles is 2092. The Isles belong to the Duchy of Cornwall but have been leased to Mr Dorrien-Smith. A large business is done in the sending of flowers and early potatoes thence to London. There is a good pier and harbour. (pp. 57, 59, 96.)

Saltash (3357), built on a steep slope falling to the Hamoaze, returned two members to Parliament till disfranchised in 1832. The Royal Albert Bridge here crosses the Tamar, erected in 1857–9, carrying the trains of the G.W.R. into Cornwall. Saltash has a station. It was formerly the great place of crossing from Devon to Cornwall and *vice versa* by ferry boats. (pp. 16, 132, 133, 139, 147.)

Stratton and **Bude** (2308). Stratton is a small market and union town, and Bude its modern prosperous daughter is two miles off, the terminus of a branch of the L. & S.W.R. whereas

Stratton has no station. It is a sleepy, decaying place; it possesses a fine church in which is the altar tomb of a Crusader, and another to Sir John Arundell of Trerice. On the other hand Bude is a growing and thriving watering-place, and possesses a small harbour. (pp. 32, 33, 57, 103, 120, 133, 135, 136, 137, 139.)

Truro (2215, but with Kenwyn 5836, and with St Clement 3283 more). Truro is a city, the seat of a bishop, a municipal

Truro

borough, head of an union and county court district, and a port. The city is situated in a valley and comprises the parishes of St Mary, Kenwyn and St Clement. At high water the tide forms a fine lake two miles in length reaching from Truro quay to Malpas, and the channel is of sufficient depth to be navigable for vessels of upwards of 70 tons. Truro was formerly one of the towns having the privilege of stamping tin, and is the seat of the Stannary Court. In 1876 Cornwall was cut off from Devon for ecclesiastical purposes and Truro constituted the seat of the bishop, and in 1880 the present cathedral was begun.   Truro possesses a Passmore Edwards Free Library, and it is the seat of the Royal Institution of Cornwall which has here a museum and a library.   (pp. 38, 66, 96, 104, 123, 124, 132, 135, 137, 139, 144, 146.)

Tywardreath (2215) is a little town where once was a Benedictine priory.   It has a coastguard station and a police station.   (p. 48.)

Wadebridge (2186), a market town on the river Camel, partly in the parish of St Breock and partly in that of Egloshayle.   It has a station on the L. & S.W.R. whence also runs a branch to Bodmin and another to Padstow.   A stone bridge originally of 17 but now of 15 arches, erected in 1485, connects Wadebridge with Egloshayle.   The chief trade of the town consists in the exportation of granite, china-clay, and agricultural produce, and the principal imports are coal and timber.   (pp. 14, 15, 135.)

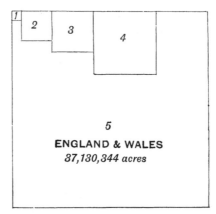

Fig. 1. The Area of Cornwall (2) 886,384 acres as compared with
that of Rutland (1) 97,087 acres, Devon (3) 1,666,839 acres
and Yorkshire (4) 3,870,038 acres

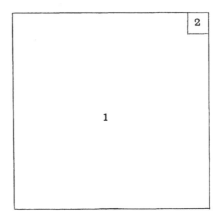

Fig. 2. Population (318,591) of Cornwall compared
with that of England and Wales in 1901

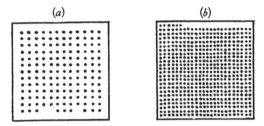

Population of England and Wales (*a*) in 1801, 153;
(*b*) in 1901, 558 per square mile

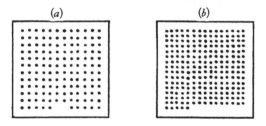

Fig. 3. Population of Cornwall (*a*) in 1801, 142;
(*b*) in 1901, 230 per square mile

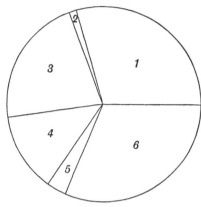

Fig. 4. Diagram showing Proportionate Land Cultivation in Cornwall (1) Permanent Pasture 252,194; (2) Bare Fallow etc. 10,495; (3) Green Crops and Rotation Grasses 193,793; (4) Corn Crops 114,151; (5) Woodlands 32,427; to (6) Remaining Area 266,818 acres

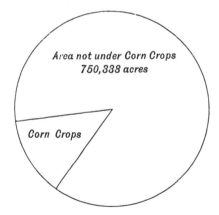

Fig. 5. The Proportionate Area of Corn Crops (114,151 acres) in Cornwall

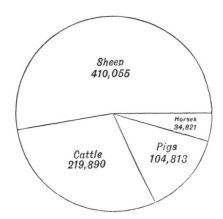

Fig. 6. Proportionate numbers of Live Stock
in Cornwall

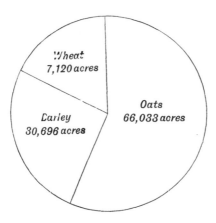

Fig. 7. Comparative Proportion of Cereals grown
in Cornwall

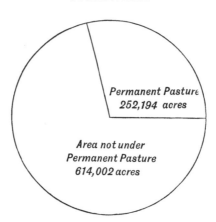

Fig. 8. The Proportionate Area of Permanent
Pasture in Cornwall

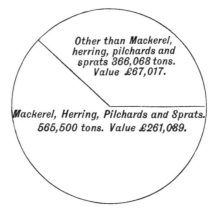

Fig. 9. Diagram showing Proportionate Weight and
Value of Fish landed in Cornish Ports

Milton Keynes UK
Ingram Content Group UK Ltd.
UKHW041520181024
449640UK00009B/86